THE
PROGRESSIVE
MACHINE

BEFORE WE CALLED IT LEAN

SECOND EDITION

STEPHEN WELLER

ISBN: 978-1-954614-39-0 hard cover
ISBN: 978-1-954614-40-6 soft cover

Library of Congress Control Number: 2021912649

Edited by: Erika Nein

Published by Warren Publishing
Charlotte, NC
www.warrenpublishing.net
Printed in the United States

For my wife, Kayti.

CONTENTS

ACKNOWLEDGMENTS

I thank my long-time friend Richard Campos. We endured much as we worked together to influence change in our organizations. Without Rich, this book would likely never have been written.

Many thanks to my colleagues in health care, including Mark Hueter, Samira Dugan, and Steven Sulkin.

A special thanks to Jesseca Chester for her artwork.

I also want to thank Vivek Ratna for keeping his promise.

Much gratitude to my father, Alton H. Crabb, for his thoroughness.

Finally, I must thank my wife. Many of her recommendations were incorporated into this book.

FOREWORD

R ome is an inspiring place to be. Walking through the Forum many years ago, a feeling of total awe struck me that has stayed with me to this day. In the middle of that vast open space, it was possible to see, at a glance, how Romans built structures on top of structures for over 2,000 years. Here were invention and persistence personified, piggybacking on those foundations that served ancient Romans so well while applying new materials and designs to suit the times.

Lean is that way too. Stephen's book wonderfully shows how Lean and its practices are built out of the rich ideas pollinated by American and Japanese innovators over 250 years of industry—spanning the Civil War through the modern age. From Ford's Model-T to Saunders's Piggly Wiggly to General Motors and Toyota's collaboration at New United Motor Manufacturing, Inc. (NUMMI), Stephen shows us how Lean was stacked brick by brick in elegant fashion.

Stephen embraced Lean principles in his extensive career within both product manufacturing and services delivery environments that were ripe for Lean implementation. His passion and courage inspired me for over a decade both through casual discussions when we were colleagues at an orthopedic medical device maker to direct tutelage as my manager in contracted hospital perioperative-based services.

I am so glad that Stephen has interwoven episodes of his Lean journey in this narrative. These are stories that have delighted me over the years because they are memorable, and especially because they reflect the true challenges of Lean practice.

"Change is easy," automaker Dr. Ferdinand Porsche once said, "improvement is far more difficult." Stephen's stories take us well beyond the ideas to what it takes to be a perceptive, and successful, Lean practitioner.

For Lean leaders and managers, this work also points to the degree of investment that is needed to make Lean practice workable across an enterprise. This is perhaps the hardest work of all and the greatest challenge to the scientific mindset that will take Lean well into the preeminent future. A future where humans and machines increasingly cohabitate and, in cases, even blend. It is in this future that the elevation of our common humanity and the spirit of Lean will become ever more important.

–RICHARD CAMPOS, PhD

INTRODUCTION TO THE
SECOND EDITION

E ach year promises new hopes and new challenges. We reach new milestones and are forced to overcome unforeseen circumstances. The year 2020 was no different.

I hoped to tell stories in 2020 about my Lean journey in both manufacturing and in health care. I wanted to connect with Lean practitioners with similar experiences to encourage them to keep persisting. That all changed when COVID-19 showed its ugly head in 2020 while I worked for a health care system.

Our hospital experienced a depletion of personal protective equipment, limited testing, and the need for new processes. The importance of my role as a Lean practitioner became clear, including working with local communities to mass vaccinate the residents. There were many sleepless nights, but my energy was fueled by the selfless sacrifice of those caring for patients inflicted with this horrible virus.

Despite the exhaustive hours spent in the hospital, I still managed to find time to write. The more I learned, the more I wrote. The more I wrote, the more I learned, so I decided to revise this book.

Persistence and progression are underlying themes within the pages ahead. These are traits I cannot help but admire. I saw these traits in many individuals who navigated the COVID-19 storm. I hope this book motivates you to persist and progress too.

INTRODUCTION

I was spoiled when I started my manufacturing career because our facility already embraced the Lean methodology. I believed our Lean manufacturing practices were the standard across all manufacturing facilities around the globe. I thought Lean was common sense, but I discovered that I was wrong when I accepted a position as plant Lean manager with another manufacturing company. We engaged a consultant to help us improve our Lean culture. The consultant provided us with a thorough assessment program, for which I had all the questions. During the first year, we followed a continuous process of self-evaluation using the consultant's assessment program to measure the progress and effectiveness of our Lean journey. We weren't doing well.

We had a promising 5S program, which was our housekeeping methodology that made the work environment easier to do our jobs. The disciplines of our 5S program were sort, straighten, shine, standardize, and sustain. The program provided a strong foundation to build a Lean system and was embraced by our leadership because they enjoyed giving visitors tours of our facility. Despite our success with 5S, we accomplished little in the areas of production flow and continuous improvement.

One of my mentors, who implemented a successful Lean program in France, visited me in the United States. I asked him for advice. He said, "You need to just do it." We needed to focus on *kaizen*, or continuous improvement, an action oriented approach, versus evaluation, a passive-oriented approach. I followed his advice, and *kaizen* events commenced immediately.

We developed our first Lean training module to facilitate *kaizen* events, which gave me an appreciation for the challenges of change management. My team quickly learned the importance of our roles as teachers, coaches, and facilitators at this stage in our Lean journey.

We persisted with *kaizen* events, and we celebrated our successes. Sustaining improvements remained a concern for me because I found previous attempts to undergo a Lean journey throughout the factory. On several occasions I would flip a corkboard around and find old value stream maps. There was every indication our work would also end up on the back of a corkboard, only to be discovered a decade later as another failed attempt.

During one of our *kaizen* events, I updated a senior leader on a Total Productive Maintenance (TPM) initiative. We increased the cleaning cycles of a machine, which in turn drastically increased production time and machine quality. He became quite upset because he believed it was impossible to increase both production and cleaning-cycle times, and he was in no mood to listen to any ridiculous explanation on how this feat was accomplished.

I stormed out of his office and bumped into the plant manager. I expressed my frustration with leadership to him—every time I turned around, a senior leader was swearing or yelling despite the support of the middle management. The next day, it was over. Nobody yelled or swore anymore. They became receptive to the new ideas Lean had to offer. I wondered what conversations had taken place the previous night.

My senior leaders struggled because many felt Lean was a fad. I created a presentation that showed the key innovators of Lean and their contributions. Leadership's eyes widened when they learned that Lean had origins dating back over two centuries. The presentation helped the leaders understand that Lean was proven to work, but success required persistence.

The senior leader who expressed his adamant concern about the cleaning cycles on his machine invited me back into his office and asked me how we increased both the cleaning cycles and the machine uptime. The cleaning cycle did not require monitoring, so the operator recommended that he activate the cleaning cycle each time he went on break. By the time he returned, the cleaning cycle was complete, and production could resume. Instead of two cleaning cycles each shift with three breaks, or ninety minutes, of downtime, we now had three cleaning cycles during the three breaks, which reduced the downtime to sixty minutes. We increased the cleaning cycles from thirty minutes per shift to forty-five minutes, which led to better machine performance.

The respective senior leader asked how he could learn more. I pulled up a website on the Toyota Production System (TPS) overview program and pointed out that the next class was scheduled within a couple weeks. Without hesitation, he said, "Let's do it. Just the two of us."

We flew to Charlotte, North Carolina, but could not proceed farther because the snow grounded all flights. We rented a car and, with white knuckles, drove the rest of the way to Georgetown, Kentucky.

We arrived that morning a couple minutes late. The senior leader was impressed with the instruction and the simulation. He began networking with the Toyota trainers, and within a couple months, Toyota trainers were in our facility teaching us about TPS. This marked the start of our true Lean journey.

Driving in a snowstorm to Georgetown, KY

Negative perceptions about Lean still exist today. A friend of mine with over twenty-five years of Lean experience expressed his concern about the commercialization of Lean. Hopefully, this book helps remove some negative perceptions.

I invite you to explore the evolution of ideas that make up Lean systems. We will attempt to credit those individuals who provided important innovations. Those individuals were passionate, persistent, and sought performance. All were action oriented.

This is their story.

PART I

1760–1940: THE INDUSTRIAL REVOLUTION TO WORLD WAR II

Have you ever read a book or an article about Lean, but the definition was not clearly stated or, crazier yet, simply not given? Let's provide one in the context of this book:

Lean is a system that empowers stakeholders
to identify and eliminate that which is nonvalue-added
with the goal to reduce lead time.

This definition stems from a quote by the one credited with the TPS, Taiichi Ohno, in which he stated:

"All we are doing is looking at the timeline,
from the moment the customer gives us an order to
the point when we collect the cash. And we are reducing the
timeline by reducing the non-value adding wastes."

Lean focuses on continuous improvement and incremental changes. Often these changes stem from simple observations that can be changed on the spot. Other times these changes lead to disruptive concepts that transcend industries.

The five people discussed in Part I clearly introduced both disruptive and powerful ideas that contributed to both Lean production and Lean management systems. Lean production systems are the methods and functions used to convert an input into a desired output. A Lean management system manages the Lean production system. For sake of simplicity, we will consolidate both the production and management system and call them Lean. They did not.

The world was transformed from 1760 to 1940. Change accelerated at an unprecedented rate. This was the time in which five early contributors to Lean lived. They developed new ways to improve efficiency with superior quality, whether it was in manufacturing or how we simply shopped for groceries.

These five men proved new ways to increase throughput with high levels of productivity to minimize costs. Sometimes motivated by profit, they were creative and persistent.

These five men never sat in a room together and said, "Hey, let's create a Lean system!" Rather, they demonstrated a constancy of purpose to create value. This constancy of purpose still defines Lean today.

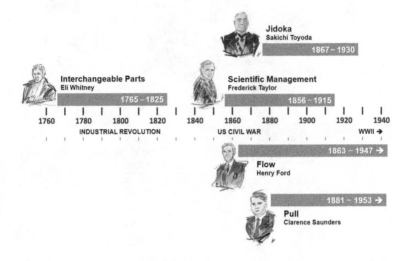

We entered a world where craftsmanship and apprenticeships were common, and the horse was a standard form of transportation. In the United States, slavery was on the rise in the South, and the Industrial Revolution was taking a foothold, quickly introducing new jobs. During this period, the Civil War would be fought in the United States, man would learn to fly, and the automobile would proliferate.

Let's begin this story before the turn of the nineteenth century with Eli Whitney.

CHAPTER 1

ELI WHITNEY
(DECEMBER 8, 1765–JANUARY 8, 1825)

*"One of my primary objects is to form the
tools so the tools themselves shall fashion the
work and give to every part its just proportion."*
—Eli Whitney

INTERCHANGEABLE PARTS

Eli Whitney was born before the American Revolution and lived his entire life during the Industrial Revolution (1760–1840). He saw the birth of the United States of America, including George Washington's presidency from 1789 to 1797, and the prominence of slavery in the South.

Whitney was the eldest of four children.[1] He attended Yale College from 1789 to 1792, but after graduation, Whitney was unemployed.[2] He originally considered a career as a lawyer, but he needed money quickly to pay off his debts. He moved south to Savannah, Georgia, where he found a job as a tutor working for Catherine Greene.[3]

COTTON GIN

Cotton was a major industry in the South but was unprofitable, even with slave labor. It was difficult to separate the seeds from the cotton. Catherine urged Whitney to develop a way to separate seeds from the cotton more efficiently. Whitney believed if he could invent a mechanism to accomplish this goal, his financial troubles could be resolved. Whitney took on the challenge, and in 1793 he invented the cotton gin.

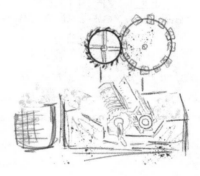

The cotton gin became popular very quickly. In the South, it made cotton production and slavery profitable. By the mid-nineteenth century, approximately 75 percent of the world's cotton originated from America. Whitney saw few profits from his invention. Even though the cotton gin was patented, other planters built variations of their own. Whitney sought a legal solution through the courts, but this became very costly. Any amounts awarded to him were slim, and by 1797 Whitney was $4,000 in debt.[4] He sought wealth elsewhere.

INTERCHANGEABLE PARTS

Eli Whitney embarked on a new opportunity to build wealth when he contracted with the US government to manufacture muskets. He pioneered a method to produce muskets that promised improvements in both production and quality. This method differentiated him from traditional musket manufacturing because Whitney built machines to produce parts that were interchangeable.[5]

At this time, much of production was considered craftsman, or handcraft, production. For example, if a craftsman set out to produce a musket, every component could only be used for that particular musket. If a component needed replacing, one could not simply use the same component from another musket. The parts were not identical nor interchangeable. Whitney's work changed this.

Whitney's machines produced parts individually. Those parts were practically identical to each other and could be used interchangeably with any other musket of the same design. Machines were dedicated to producing individual parts repeatedly, requiring fewer complexities to learn by any individual worker in the manufacturing process. Workers with less skill could still produce great quantities at lower costs.[6]

Whitney could take several muskets and disassemble them, mix up the parts, and reassemble fully functional muskets. Imagine the awe of those observing such a demonstration at the time. Many years later, Henry Ford witnessed a similar demonstration with three automobiles.

Whitney claimed he invented interchangeable parts, but it might be more accurate to say he likely reinvented the concept. There is substantial evidence to this point. In 1720, Christopher Polhem of Sweden used machinery to produce gears for watches so interchangeability could be ensured.[7] There's also evidence of interchangeable parts that date back to the First Punic War (264–241 BC), during which the Carthaginian ships had interchangeable components, and thus warships could be mass produced.[8]

Without a doubt, though, Eli Whitney pioneered the practical application of interchangeable parts, which he introduced to the American industrial system.[9] In fact, this concept would be a major enabler to the manufacturing industry in the North.

INTERCHANGEABLE WORKERS

The manufactured parts were not the only elements in the manufacturing process that were interchangeable; so were the workers. Workers could be trained quicker than a craftsman, and each could be used at multiple production stations. Worker productivity improved.

Interchangeable parts created a need for occupational specialization. Colonel Roswell Lee, the superintendent of the Massachusetts Springfield Armory in 1815, worked to specialize labor in support of Whitney's innovation. Within the Springfield facility, occupational specialties increased from thirty-six in 1815 to over 100 by 1825.[10] Specialization of labor could be credited in part because the interchangeability concept depended on process simplification and standardization.

I worked many years in health care helping to improve operating-room (OR) performance. Surgeons usually want the following: their choice of surgical supplies and instruments, shorter wait times between surgeries, and the same OR staff. In many cases, the surgical volume does not justify a dedicated surgical team.

Supplies and instruments are usually recorded on a surgeon's preference card, often in the form of a picklist. Many times, there is a wide variance in the supplies picked between surgeons performing the exact same procedure. One surgeon may list twice as many items than another surgeon. Of all items kept on hand for a specific procedure, as much as half is used by only one surgeon.

Why is this relevant? We found, in many cases, very close correlations to both the number of items on the picklist and the downtime of the OR between surgeries. In some cases, surgeon-to-surgeon differences had statistical correlations with surgery durations.

There is an alignment between the three surgeon priorities. First, if we present the total list of items found on all cards for surgeons to pick from, they may find the list to be extensive. The more they pick what other surgeons use to replace their own, the more standardization is achieved. The more items removed from the preferences, the fewer items staff handles, which helps to remove motion, thus potentially lessening time between cases and ultimately reducing surgeon wait time. Finally, the more surgeons standardize with each other, the fewer exceptions must be addressed by surgical teams. Less complexity between the different physicians helps OR teams develop greater levels of expertise, and thus they can support a greater number of surgeons with improved efficiency.

Reduction of line items is a way to simplify the cards, whereas the use of same items results in standardization. Simplification and standardization help promote the idea of interchangeable parts between each surgeon and empower interchangeable staff. Surgeons have the instruments and supplies of their choice, wait times are reduced, and surgical teams can support each surgeon more effectively. Along with other factors, preference card standardization and simplification help to improve OR efficiency, cost, time durations, and clinical and safety performance. For example, door openings during surgery can be reduced because staff do not need to leave the OR often for supplies.

THE NORTH AND THE SOUTH

Whitney's cotton gin helped increase slavery in the South. On the other hand, his pioneering efforts of the interchangeability of parts impacted the industrialization of the North. This industrial strength, along with many other factors, helped the North achieve victory over the South during the American Civil War (1861–1865). One may suggest that the disparity of these two extremes may have balanced itself with the American Civil War.

VARIATION REDUCTION

Interchangeability of parts and the specialization of labor were early attempts to address variation within a production system and drive improvements both in production and quality. In Lean systems, variation is commonly termed *mura*, a Japanese word. In his book *Gemba Walks*, Jim Womack defined *mura* as "unevenness in demand not caused by the end customer."[11] Womack's definition translates closer to the true meaning of *mura* than variation.

Standardizing parts to ensure interchangeability helped to reduce variation in a very direct way. Interchangeable parts influenced specialization of labor, something that helped enable both Frederick Taylor's scientific management and the program in World War II known as Training Within Industry (TWI). Interchangeability was a foundation from which Lean systems would build upon, most notably by Henry Ford in the creation of the assembly line.

CHAPTER 2

FREDERICK TAYLOR
(MARCH 20, 1856–MARCH 21, 1915)

"What we are looking for, however, is the ready-made, competent man; the man whom some one else has trained. It is only when we fully realize that our duty, as well as our opportunity, lies in systematically cooperating to train and to make this competent man, instead of in hunting for a man some one else has trained, that we shall be on the road to national efficiency."[1]
—FREDERICK TAYLOR

SCIENTIFIC MANAGEMENT

Frederick Winslow Taylor was born in Germantown, Pennsylvania,[2] before the American Civil War and died during World War I, about two years before the United States declared war on Germany. Early in his career, Taylor spent time within manufacturing facilities, where he observed significant inefficiencies in the way workers performed their tasks. Taylor developed a belief that a person's work in a manufacturing facility could be designed. He began working on an objective, fact-based system to determine performance standards for workers.

Taylor employed the use of time studies and was possibly the first to use a stopwatch to evaluate work for analytical and constructive purposes.[3] Tasks were evaluated to eliminate any unnecessary movements and how to best organize the rest. The determination of unneeded motion was a precursor to the Lean concept of waste elimination.

Taylor focused on the effective combination of work steps to complete the work but not necessarily the evaluation of individual tasks to determine if that activity created value or not. This approach drew criticism from Henry Ford.

Let's assume that it is my job to write my name on a dry-erase board. My habit is to walk to the board, perform a happy dance, then walk across the room to retrieve a dry-erase marker, walk back to the board, write my name on the board, then return the marker. Someone using Taylor's approach would note each step, time it, possibly move the marker closer to the board, and standardize this new process. However, that person may not question the happy dance. Henry Ford most likely would question the value of the happy dance. When we think of manufacturing and other processes, we often perform activities that add no more value than the happy dance, and standardizing the expectation for such an unnecessary activity is nonvalue-added.

Because these nonvalue-added tasks are not removed, the worker, in essence, must work harder.[4]

Taylor used time studies to determine realistic goals for workers that could in turn be used for setting a piece rate for each worker. Incentives were aligned to production. Workers were paid based on individual production, not necessarily for hours worked.[5]

In 1911, Taylor published *The Principles of Scientific Management.* He hoped his book would encourage others to seek a systematic management approach to remedy inefficiencies, and to prove that science—using clearly defined laws, rules, and principles—would create the best form of management.[6] There are four principles in Taylor's scientific management:

1. Managers develop a science for each element of a person's work.
2. Managers scientifically select and then train, teach, and develop the workman.
3. Managers cooperate with the worker(s) to ensure work is in accordance with the science developed for that work.
4. Managers take over all tasks better suited for the manager, creating an equal division of work between the workman and the manager.[7]

Taylor noted on several occasions American management sought to find the extraordinary worker versus designing work so the common worker could be efficient. This could theoretically be a source of the following:

Question: Who should you hire, an expert or someone new with potential?

Answer: It depends. How good are your processes? If you have poor processes or no processes, you may need the expensive expert. If you have good processes, then you may not need the expensive expert.

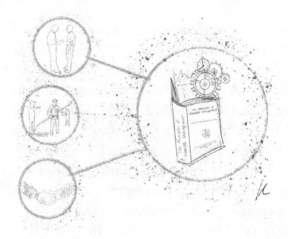

Taylor took some heat for his statements about labeling workers as either first-class men or not. Taylor believed workers needed to have their abilities match their jobs. If appropriately matched and the worker chose to work, then the worker would be considered a first-class man. In his testimony to a congressional special committee, Taylor stated that only those workers who were "perfectly well able to do the job but won't do it" were not considered to be first-class men.[8]

> As a manufacturing leader, I had to motivate workers to produce work to the best of their ability and ensure capabilities met production demand. If a person's employment was terminated, it was typically for some form of negligence or attendance violation.
>
> I do not recall ever disciplining anyone for a quality or production deficiency because I always felt I held some responsibility, hence my adaptation of Taylor's fourth principle.

> A person who tried did not need disciplinary action. Mentorship counseling was a tool we embraced to address performance concerns. We involved the worker in developing a plan to reach a respective goal. Positive incentives and resources were provided. If the person could not meet the goal, the person was likely reassigned to a different position.
>
> I had an employee whose job was to manually buff products. The worker's quality was good, but production volumes were low. Together we devised a performance-improvement plan.
>
> We reviewed progress regularly and production improved. The employee was promoted. Empowered by the improvement, the employee became a mentor and commonly said to others, "If I can do it, so can you. I'll help you."

Taylor advocated working smarter, not harder.[9] By studying the work, using data, putting measures in place, and the provisioning of resources, management could drive efficiencies.

Time and motion studies were used to break down a job into simple and separate tasks that could be performed repetitively by all workers without deviation.[10] Taylor believed that work standards and rules could define and control the performance by workers, and efficiency was able to be gained by reducing the complexity within the job.[11] Some labeled Taylor as the father of the time study, industrial engineering, and scientific management.[12]

Taylor influenced others such as Frank and Lillian Gilbreth and Henry Laurence Gantt. Frank and Lillian became the parents of motion studies.[13] Motion studies focused on a worker's movements such as the amount of reaching, walking, and other physical actions that occur in the completion of a task. A motion study may have highlighted that a worker consistently reached for a tool. The tool may then be repositioned to reduce the amount of reaching by the worker.

Henry Gantt emphasized the criticality of leadership in the workplace. Gantt developed the Gantt chart, a concept of planning based on time versus product quantities. In essence, he created a chart that evaluated work based on comparing actual performance to planned performance. Many people today are familiar with the use of Gantt charts in project management, but its origins were in planning manufacturing operations and scheduling.

Scientific management made its way to Japan via a translation of Taylor's publication *The Principles of Scientific Management*. Shigeo Shingo, who may have been the world's first Lean consultant, found a copy of Taylor's book in 1931. Shingo meditated deeply on Taylor's statement that high-paid workers could profitably produce inexpensive goods.[14] Shingo introduced quality and production concepts that were combined to achieve financially effective solutions by using Taylor's thinking to design efficiencies into work.

About a half century later, standardized work became an integral aspect of the TPS to drive a truer form of scientific management as opposed to Taylor's, which seemed to have more of a command and control feel.[15] Toyota's standardized work was based on understanding the work content. Toyota designed the work to meet the demand without overburdening resources. Overburdening was often referred to as the Japanese term *muri*, which was also translated as hard to do.[16] *Muri* could be caused by inadequate design of work, something Taylor addressed directly.

Standardized work was a way to organize work based on demand. It could show how a person performed his work as it related to the automated processes of a machine to prevent how much time was lost by improper sequencing. Periods of lower production demand may have required fewer workers performing a broader range of tasks, whereas periods of high demand may have required more workers performing fewer tasks. The work was balanced between the workers. Note how science was used to design work.

The science of timing and evaluating work steps, designing work based on data, and the use of measurements had its origins with Taylor.

CHAPTER 3

HENRY FORD
(JULY 30, 1863–APRIL 7, 1947)

*"There is a difference in a man working hard and hard work.
A man working hard will produce something, whereas
hard work is the least productive sort of labour."[1]*
—Henry Ford

FLOW

Henry Ford was widely known as the founder of the Ford Motor Company who brought manufacturing and assembly production to new levels of efficiency using innovative techniques that culminated in the creation of the assembly line. The success of the assembly line stemmed from a combination of advancing both work design and the use of interchangeable parts and workers. Much like how water flows through a riverbed, Ford's vehicles flowed through the production line.

Flow is a concept that describes how the product, or other forms of value, move throughout the transformation process. Stopping the flow can come in the form of inventory (stationary products) and waiting (stopped workers). The slowest process step, or a constraint, is the maximum speed of the entire production process. Constraints come in many forms.

When thinking about flow and its importance in a production process, let's use the analogy of a river. If a riverbed is smooth and relatively the same width and depth throughout its entire pathway, then water can flow rapidly. We may not even be aware of how fast the water is flowing unless we put a finger in it.

Choppy flow   *Smooth flow*

On the other hand, if there are many boulders throughout the riverbed, the flow of the water is obstructed, causing white water. The speed of the water is easily noticeable. In the same way, a manufacturing process that is not leveled and has many obstacles will also be disrupted.

Several influences in Ford's production program created effective flow of products through his assembly lines, most notably the concept of interchangeability. There was also evidence of scientific management.

INTERCHANGEABILITY

Ford's introduction of interchangeable parts to the automobile industry may have stemmed from Henry M. Leland.

Leland previously worked at the Springfield Armory, where Eli Whitney's concept of interchangeable parts prevailed. Much like the exercise in which Whitney mixed up musket parts and reassembled them, Leland did the same, except he was doing it with automobiles. In 1906 London, Leland disassembled three Cadillacs, mixed up the parts, and reassembled them back together. To Ford, this was proof that he could use the concept of interchangeable parts in the manufacturing of automobiles.[2]

To manufacture parts that were interchangeable, standardization was needed, and machines were critical to ensure that standardization.[3] In this way, the armory production systems, influenced by the interchangeability of parts, became a precursor to Ford's manufacturing system.[4]

SCIENTIFIC MANAGEMENT

There also were efforts to employ the use of scientific management in automobile manufacturing, with the work of Carl G. Barth (1860–1939) being noteworthy.

Barth was employed by the Franklin Automobile Company between 1908 and 1912. Parts were brought to a workstation, where a vehicle was assembled. Barth's work improved car production from 100 per month to forty-five per day, reduced labor turnover from just over 42 percent to less than 50 percent, and increased wages by 90 percent.[5]

Barth's results were impressive, but Ford's assembly line produced scalable results. Barth assembled a vehicle in a single, stationary location. Ford physically moved the vehicle throughout the production process. The latter had numerous benefits, most important of which was the reduction in movement of labor and material by bringing the car to each step of the assembly process. Ford's team designed work around the assembly line.

Ford was critical of Taylor and his scientific management because not all work in Taylor's model created value. Ford appeared to want nothing to do with the scientific management community, and the feelings may have been mutual. Nobody closely associated with scientific management performed any work at the Ford Motor Company. Ford denied any dependence on scientific management, but it was likely that some of his engineers used some of Taylor's ideas in their work.[6] Ford replaced Taylor's piece-rate system with an hourly wage system, and he gave production process controls to his engineers.[7]

Frank Gilbreth, an advocate for scientific management, visited a Ford assembly plant. Gilbreth noted several ergonomic issues such as low benches from which workers constantly stretched to perform their work. According to Gilbreth, management claimed workers did not mind such conditions and that they simply adjusted.[8]

If we went back in time and asked Mr. Ford if he was influenced by the idea of the interchangeability of parts in the automobile manufacturing process, it is likely he would have responded with a resounding *yes*. On the other hand, if we asked him about the influence of scientific management, it is likely he would have responded with an equally resounding *no*. Evidence, though, suggests that Ford's production system was influenced by scientific management, whether he would admit it or not.

Scientific management could be characterized as: production work separated from planning work, standardization of work, reducing work time required, and continuous improvement.[9] The use of science versus common sense to design work demonstrated that Ford, knowingly or unknowingly, applied scientific management concepts successfully to his assembly processes. The design of work to ensure flow by the assembly line propagated the idea that Ford's team designed quality into the manufacturing process. For this reason, Ford was a practitioner of scientific management.[10]

ASSEMBLY LINE: A NEW APPROACH

In 1910, prior to Ford's assembly line, individual Ford workers only produced 6.73 automobiles each on average (2,773 workers; 18,664 automobiles) with a $950 price tag on the Model T Touring Car. Four years later in 1914, each worker produced almost triple the amount, averaging 19.28 automobiles (12,880 workers; 248,307 automobiles) with a $490 price tag on the Model T Touring Car—about half that of the 1910 price tag.[11] Designing a production system to flow was certainly a game changer, especially in the US market.

Assembly line

An assembly line on a conveyor belt helped establish the pace of production. Work distribution between workers needed to be designed to complete all tasks without having to stop the line. Conveyor belts allowed workers to perform production on the assembly line while products could be placed near the worker. In short, the worker moved little, giving workers the opportunity to use most of their time performing value-added tasks such as assembly.

While working for a medical device company, I was asked to go to a hospital to review the sterilization process and determine why surgical instrument trays were not ready for surgery.

I had never worked in a hospital before, and I knew nothing about sterilization of instrument trays. Up until that time, I avoided hospitals like the plague because of a bad experience when I was twelve years of age. I cut my foot severely and was taken in for stitches. The doctor only partially numbed the area and proceeded to stitch. I wanted to cry out with each stitch. My mother, who was a nurse at the same hospital, told me not to act like a baby when I twitched with each stitch. I felt like Charles Schultz's characters, who often screamed, "AAUGH!"

Anyway, I was instructed to arrive at the hospital on Tuesday morning and to brief the leadership of both the hospital's perioperative services department and my organization as to the cause of the delays by Friday morning.

I interviewed workers, observed the process, prepared motion maps of the staff performing their work in the sterilization process department (SPD), created transportation maps of the flow of instruments, and prepared a capacity analysis. By midafternoon, all my calculations suggested there was plenty of capacity to meet the demand. The only problematic areas in the process were the elevators used to transport the instruments to and from the surgical floor.

Not all elevators were in commission; those that remained delivered surgical carts from the OR sporadically. Flow was not constant. I needed to understand more about how significant this unleveled flow was, so I remained during the afternoon and night shifts.

Soon into the afternoon shift, the elevators deposited carts more frequently. In fact, the elevators did not stop. Before long, there was a sea of dirty surgical carts with instrument trays. There was no way to process all carts in time for the cases needed the next day.

I asked to go into the surgical department to observe what happened with the dirty surgical carts after surgery.

Carts with instrument trays were stationed by the elevator, waiting to be returned downstairs for cleaning and sterilization. The elevator had a sensor to identify the presence of a cart. Upon sensing the cart, the elevator opened, pulled the cart into the elevator, and transported it to SPD. The carts were stationed on a ramp built into the floor so gravity could move the cart into position near the sensor.

The sensor did not always work. Also, the ramp could only hold two carts at a time, so excess carts were stored elsewhere. The carts sat around for extended periods of time as they piled up.

My observations were reported on Friday morning as promised. The hospital agreed with my report and said they would address this issue. Awkwardly, three years later, I revisited the hospital, and the same issue still existed.

The flow was disrupted.

Nowadays, when evaluating surgical flow, one of the first questions I seek to answer is whether the SPD is on the same floor as the OR because of the significant loss of time associated with waiting for elevators.

Imagine my pleasant surprise when reading Section 5.2.3 in the DoD Space Planning Criteria dated July 1, 2017, which stated, "Sterile processing should be located on the same level and contiguous [sharing a common border; touching] to the surgical suite." I was not the only one with a similar viewpoint.

WASTE REDUCTION

Ford and his team were early pioneers who actively focused on the elimination of waste, or nonvalue-added activities, in their assembly lines and production processes. Waste drove up the cost of production, which could be found in the form of overproduction, waiting, conveyance, excess processing, inventory, motion, and correction of defects. Waste also drastically increased a production system's lead time.

Ford wrote about waste reduction in his 1926 book *Today and Tomorrow*. He mentioned that having too much raw material (inventory) was a waste, and it drove up prices and lowered wages.[12] According to Ford, wasted time could not be recovered; a worker waiting instead of producing was time lost.[13] Ford acknowledged that having too little inventory could also be very harmful in the event there was a defect, which could potentially stop production. He also pointed out the need to evaluate the ease of transportation of the material, making a connection between transportation, inventory and defects.[14] In Chapter 10 of his book, Ford focused on transportation, emphasizing the need to evaluate the cost of transportation throughout the assembly process.[15] Why was this significant? Ford recognized that non-value-added work negatively impacted the success and profitability of his company.

FORD'S VIEWS

Henry Ford believed that real businesses started with production and that industry was made up of ideas, labor, and management.[16] These qualities were sources of money.[17] He stated three principles:

1. "Do the job in the most direct fashion."
2. See that the worker is actively employed while working.
3. Maintain the machines.[18]

Ford was also an early pioneer of ongoing improvement to increase quality and reduce prices.[19] He believed a standardized method would be the best of all methods and that a standardized method should be improved upon. The best method had the characteristics of simplicity and the elimination of nonvalue-added work. The best method also correlated to improved safety and quality as production improved and costs were reduced.

ASSEMBLY LINE IMPACT

The assembly line was an amalgam of several designed manufacturing processes that produced large volumes of vehicles at a cheaper cost, thus lowering the price to the domestic consumer. Ford created a production system that began to maximize economies of scale, or producing great numbers efficiently, resulting in massive quantities of vehicles.

Mass production was and is commonly referred to as "Fordism." Ford could mass-produce profitably because of the immense market in the United States for his automobiles. He had easy access to capital, regulations were liberal, and there was a plethora of unskilled workers. He focused on the complete and consistent interchangeability of parts and kept the process simple, thereby promoting the interchangeability of people.[20] Ford quickly converted an unskilled worker to a skilled worker because jobs were relatively simplistic.

The TPS concept was heavily influenced by Ford's assembly process. Toyota's Taiichi Ohno had great respect for Ford, stating, "Henry Ford was able to mine iron ore on Monday and, using that very same iron ore, produce a car coming off the assembly line on Thursday afternoon."[21]

CHAPTER 4

CLARENCE SAUNDERS
(AUGUST 9, 1881–OCTOBER 14, 1953)

*"One hundred people can wait on themselves
at Piggly Wiggly ... every forty-eight seconds
a customer leaves Piggly Wiggly with her purchase."*[1]
—CLARENCE SAUNDERS

PULL

Lean employs two important concepts: flow, as noted in the previous chapter with Henry Ford, and pull. Clarence Saunders, like Henry Ford, sought a new approach to achieve profitability for his business. Ford's system of flow worked because of the high demand for his product, and much efficiency was gained with his assembly line as a focal point. Consequently, the Ford production system required a considerable amount of space and inventory, thus came the need for pull.

Clarence Saunders grew up in the rural areas of Clarksville, Tennessee.[2] His family was not wealthy at the time of his birth, but his ambition, persistence, and hard work helped to change his circumstances.

Saunders founded the Piggly Wiggly grocery chain. His grocery customers wanted a diverse selection of products in a central location. To provide a selection that customers wanted was not profitable if there was an exhaustive amount of space used to house large inventories of diverse products. Saunders needed a different approach. This approach became one of the earliest forms of pull.

BEFORE PIGGLY WIGGLY

The grocery shopping experience before Piggly Wiggly entailed submitting an order to a shop clerk and waiting for the clerk to pull and give the customer the ordered product.[3] There could only be one clerk per customer throughout the entire shopping process, so costs were inherently excessive. Saunders noted that if shoppers tried to serve themselves at the shop, the clerk would intervene. Saunders pointed out that clerks were often very inconsiderate to customers with such statements as, "Take what is given you."[4]

PIGGLY WIGGLY

Manufacturing in the food industry evolved, and food packaging made it easier for consumers to prepare meals. Manufacturing preservation techniques such as canning allowed consumers to eat foods not in season.[5] These new methods were exploited by Saunders in the creation of a self-serve grocery store.

Saunders opened his first Piggly Wiggly market on September 6, 1916.[6] He took advantage of the new packaging and preservation methods to find new ways to display, store, and market products. Most importantly, he created a new grocery business model in which customers shopped and pulled products from the shelves with little to no assistance from clerks. This allowed a greater number of customers to be serviced while reducing the cost from excess labor.

Piggly Wiggly became the first successful supermarket because the stores differentiated themselves from other grocers.[7] Prices were on each item, national brands were advertised, checkout stands were used, produce was preserved through refrigeration, and employees wore uniforms to help improve the safe handling of food.[8]

The shopkeeper's role changed because customers were now pulling their own items. It was here that the pull concept was born. Each item had a designated space with stocking levels determined and identified. Shopkeepers checked the shelves regularly. Empty shelves triggered replenishment from the storeroom.[9] Regular replenishments made it possible for Saunders to keep a large variety of products in smaller spaces available for customers to select from.

Item pulled   *Item replenished*

The pull concept is the opposite of push. In a push system, large batches of inventory are created to account for variations within a production process and address just-in-case challenges. The risk of pushing product increases if an upstream process operates faster than a downstream process.

Let's assume step A takes two minutes to complete, and step B takes three minutes. In a push system, step A keeps producing without regard for step B and thereby increases the inventory sitting at step B. A pull concept minimizes the levels of inventory between steps A and B. Step A only produces whatever step B pulls from the inventory. In a supermarket, when the customer pulls an item, the clerk replenishes it. The same concept applies with supermarkets in production systems.

As an inexperienced night production supervisor within a manufacturing plant, I was requested by another department to triple the production of one product. I was on the job for only a short period, transitioning from a different industry, so like a chump, I obliged. There were two major problems with this decision. First, I produced more than needed, hence I pushed too much product. Second, I disrupted the flow of other products and brought some production lines to a standstill.

My boss arranged for me to meet one of the shift leaders who had much experience with pull systems. The shift leader led me to the shop floor supermarket and explained how it worked. When a product was used, a card attached to the product container was sent back to the previous operation, signaling the need for replenishment. This helped ensure a regular flow of product by replenishing only what was pulled. I never made the same mistake again.

TOYOTA'S VISIT TO THE UNITED STATES

In the 1950s, Toyota personnel visited the United States with at least two goals: see American automobile manufacturing—Ford's plant in Michigan being one of them—and American supermarkets.[10] The Ford method was not practical to the Toyota visitors because of the significant use of space, inventory, and material. Toyota felt American supermarkets may hold the answer.

The Toyota team visited a supermarket, which some believed was a Piggly Wiggly, and noted that when a person pulled a drink, the drink was automatically replaced.[11] If the Toyota manufacturing and assembly processes had a supermarket of parts, then when a part was pulled, the manufacturing processes would simply replenish the part and fill the empty space, ensuring part availability and preventing unwanted inventory.

Toyota's Taiichi Ohno had already started on the pull concept prior to his visit to the United States in 1956.[12] Toyota needed a solution that kept material flowing to prevent dollars and space being tied up in inventory and inventory storage locations. The supermarket, in their eyes, was the solution, and thus it inspired Toyota's pull systems. When the customer process consumed a product, communication to the supplier process triggered a replacement.[13]

IMPACT

Piggly Wiggly's self-serve innovation would touch successful retail chains like Walmart.[14] They built upon Saunders's methods with innovations of their own. They created stores with more revenue-generating space by converting inventory space to retail space.[15]

It was difficult to prove that Saunders deliberately or personally created the replenishment techniques for Piggly Wiggly, but it is believed he may have unknowingly invented the pull principle and changed the world.[16] Toyota studied US supermarkets, which influenced the creation of other effective pull systems such as *kanbans*.[17]

Unfortunately, financial problems arose, and Saunders's persistent entrepreneurial spirit remained with him after Piggly Wiggly was lost to him. His future ventures included a store named Clarence Saunders, Sole Owner of My Name and the Keedoozle Corporation. Sadly, Saunders died after collapsing from exhaustion.

CHAPTER 5

SAKICHI TOYODA
(FEBRUARY 14, 1867–OCTOBER 30, 1930)

"There is nothing that can't be done. If you can't make something, it's because you haven't tried hard enough."[1]
−SAKICHI TOYODA

JIDOKA

Sakichi Toyoda was the son of a carpenter who lived in a small textile trade community outside Nagoya, Japan.[2], [3] While growing up, he noticed the difficulties women, including his mother, had working with their looms, so he sought practical and inventive ways to make their work easier. Concern for the workers engrained itself into Sakichi early in his life.

Sakichi proved to be a gifted inventor. He received his first patent for a loom in 1891.[4] By 1894, his looms were cheaper and performed better than competitive looms.[5] Sakichi admittedly did not draw his inspirations for his inventions through the study of mathematics or physics or by reading books and catalogues. He spent his time alone pondering ideas that ultimately led to hundreds of patents.[6] With such a strong foundation, Sakichi incorporated Toyoda Automatic Loom Works Ltd. in the late 1920s.

Castings and steel production techniques were not developed in Japan prior to 1910, so the materials used were inferior and could not easily produce parts with precision or interchangeability.[7] Poor materials contributed to the inferiority of many products that were produced in Japan. For this reason, Sakichi felt the looms produced in Japan were inferior to the looms produced in the United States and Europe.[8] He sought the help of an American engineer currently in Japan by the name of Charles Francis. From Francis, Sakichi learned about American manufacturing methods, which included both uniformity and interchangeability.[9] Interchangeability of parts was dependent on quality material to ensure uniformity and precision.

Sakichi's endeavor to build an ideal loom began in 1901, and it took twenty-five years to accomplish his goal. About the turn of the twentieth century, one of the first improvements he made was a loom that would automatically stop if any of the threads broke. *Jidoka* was born. This feature enabled other inventions and helped Sakichi accomplish his goal of a fully-automated loom.[10] This fully-automated loom was known as the Type G Automatic Loom.[11]

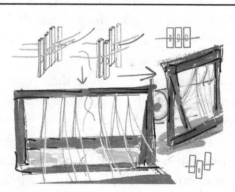

Sakichi's Jidoka[12]

The drawing above depicts Sakichi's *jidoka* concept that was incorporated into his looms.

String was fed through the rods. If the string broke, the rod, acted upon by gravity, dropped and automatically shut down the machine.

The concept was simple. The impact was profound. This was a way to build limited human intelligence into a machine in which the machine acts without waiting on human intervention.

On another note, it was assumed by some that Sakichi owned the patent for the Type G Automatic Loom, which had an automatic shuttle changer. Research conducted in 2006 revealed something else. The Type G Automatic Loom had Japan's patent number 65156, and according to paperwork associated with the patent, the patent belonged to Sakichi's son Kiichiro.[13]

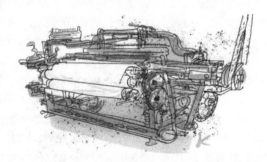

Loom

Why is this information significant? Dissenters can, and do, use inaccurate information to halt Lean journeys. Awareness of discrepancies minimizes threats to the credibility of Lean teachings. It is very difficult to regain credibility when lost.

Whatever the case, Sakichi was widely credited for creating the *jidoka* concept.

THE AUTOMOBILE IDEA SPARKED

Sakichi got the idea of an automobile company when he traveled to the United States and Europe, where he saw automobiles for the first time. He believed there were similarities between producing looms and producing automobiles. Sakichi's vision of an automobile company was achieved by his son, Kiichiro.[14]

In the 1920s, Kiichiro sold patent rights for a loom to an English company, Platt Brothers & Co Ltd.. According to some people, including Taiichi Ohno, Kiichiro used this money to conduct research on automobile designs.[15] Kiichiro, during his research, began using precision machine tools and components from General Motors (GM) and Ford to design the Model AA automobile.[16]

The question has come up as to whether Sakichi's son used the money from the Toyoda-Platt agreement of 1929 to start the Toyota Motor Company. It is agreed that patent rights were sold to Platt Brothers, but the Toyoda family may not have received the full amount because Platt Brothers claimed there were damages related to loom manufacture know-how.[17] According to the same research, a heavy tax was levied on the money. Also, a considerable portion of the money was given to Toyoda's employees as a reward for service in the past.

However the money was used, there is no doubt that work commenced to create an automobile company around the time of the sale of the patent rights to the Platt Brothers.

JIDOKA

Jidoka improved production processes and quality by addressing a flaw immediately. In essence, with this invention, Sakichi made it okay to stop production to resolve the cause of a defect.[18]

Jidoka was simply building quality into processes.[19] It was a way to create an interface between workers and machines so workers could focus on more humanistic tasks and exercise their own judgments.[20] Errors were identified by some form of technology and communicated to the operator. Defective products did not pass to the next stage of production without correction.

As a supervisor of assembly operations, I was responsible for several machines designed to help workers identify errors before the products were packaged and sent to customers. While assembling locksets, we needed to make sure that some locksets were packaged together with the exact same key cut.

Reading keys was difficult with the naked eye, so a key-cut reader was put in place on the line to assist us. The machine identified errors.

The cause had its root in batch processing. We put in place a one-piece flow system to eliminate errors produced on the assembly line so every key was perfectly sequenced. That did not prevent the keys from being delivered to us out of sequence.

The key reader is an example of *jidoka*, but it can also highlight other problems inherent in a production system. Catching a defect and preventing it from being passed on still suggests a defect is created.

Sakichi died in 1930, but his influences were profound when his son, at Sakichi's urging, birthed an automobile company that incorporated *jidoka* heavily into its production system. Sakichi's *jidoka* concept was expanded upon by others such as Shigeo Shingo in the 1950s. Instead of simply stopping production when there was an error, Toyota focused on preventing errors from occurring using such concepts as mistake proofing (*poka-yoke*) and inspecting at the source (source inspection).

Sakichi was the second person in Japan's history to be awarded the prestigious Commemorative Award from the Imperial Academy of Inventions in 1926 for inventing the automatic loom.[21] Though his invention was worthy of such an honor, it was the connection he made between loom production and automobile production that may have solidified his place in history.

PART II

World War II created a sense of urgency for the individuals covered in Part II. Whether it was to develop a program that increased national production in support of a war effort in the United States or to rebuild the war-torn country of Japan, these individuals approached their work with the idea that success was the critical imperative.

During World War II, industrial workers became soldiers. Factory floors were replaced with a less-trained workforce such as housewives, older citizens, farmers, and the like. Despite the challenges of having an untrained workforce, the United States saw a near unprecedented increase in production. This was largely credited to the work of a team deploying a program called Training Within Industry (TWI).

The United States was the only major economic power after the war, and it embarked on what could be deemed as a lazier, more complacent form of production. These production systems were indulged because the economic situation in the United States was strong. Cash flow was not tied up too long in inventory because the inventory had a high assurance of being sold. On the other hand, Japan's production systems had no such assurances. Japanese companies, such as Toyota, had to produce products with immediate demand so cash flow could be maximized.

General Douglas MacArthur commanded the US occupation of Japan from 1945 to 1952.[1] MacArthur aimed to create a free enterprise system. A few Japanese companies, called *keiretsu*, controlled about 90 percent of the industry in Japan, but General MacArthur was able to restructure the economy to break their influence.[2] With that influence broken, MacArthur was free to introduce new ideas and programs. One of the programs brought to Japan was TWI.

Douglas MacArthur

With the assistance of the United States, Japan strengthened its economic situation. Japan demonstrated its appreciation for that help when it honored General MacArthur with an elite award typically given to monarchs and government heads, the Grand Cordon of the Order of the Rising Sun with Paulownia

Flowers.[3] The verbiage within the citation included: "You helped them [Japanese people] regain their self-respect and rebuild their economic life."

Japan's financial strength swelled when the United States entered the war with Korea. Japan won contracts that supplied the US military. These contracts helped transform Japan into an industrial powerhouse.

Meanwhile, on the other side of the ocean, some Americans became concerned about US production systems. Warnings and recommendations were largely ignored by US industries. Conversely, these ideas and recommendations were embraced by Japan. As a result, quality in Japanese industries improved.

Automobile production skyrocketed in Japan. The market in Japan demanded smaller, more fuel-efficient vehicles. The US market demanded larger, fuel-inefficient vehicles. Small-car production made little financial sense for US automakers, so smaller-vehicle production was outsourced to other countries for sale in the United States. Japan was one of the small-vehicle suppliers.

Toyota helped lead the way in developing a production system that maximized cash flow. It incorporated concepts such as just-in-time production and single-minute exchange of dies. These concepts, along with others, were blended into TPS, which broke new ground in waste reduction.

By the 1970s, TPS would have all its components in place. In a short period, American manufacturing found itself considerably far behind Japanese manufacturing. The world, especially the United States, began paying attention to Japan, which, in only a couple of decades, rebirthed from the ashes of a financial crisis to a state of wealth. The superiority of quality in Japan also accelerated to industrial envy, resulting in legendary tall tales.

> I remember a tale about a Japanese business leader being asked if his company had ever produced a defect. With trodden eyes and a bowed head, he brought in a box covered by a piece of cloth. He shamefully removed the cloth and revealed his one defect for the entire world to see.

Let's discover what happened that brought such mythological levels of quality to life.

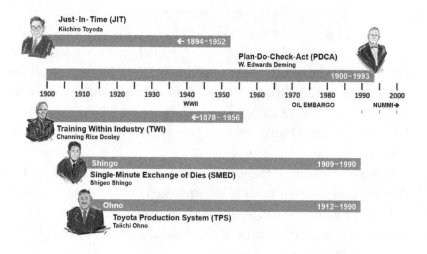

CHAPTER 6

CHANNING RICE (C. R.) DOOLEY
(APRIL 4, 1878–JUNE 25, 1956)

*"No plant management can hold a supervisor responsible
for things about which he has not been informed..."*[1]
–C. R. DOOLEY

TRAINING WITHIN INDUSTRY (TWI)

C hanning Rice (C. R.) Dooley was born on April 4, 1878*
in Rockville, Indiana. Dooley graduated from Purdue
University with a degree in electrical engineering.[2] He
was described as having "never lost his temper" as well as being
able to put a "distinctly human face" to training and learning.[3]

> *There is some discrepancy in the birthdate with at
> least one source stating Dooley was born in 1882.
> Other sources, including one that depicts Dooley's
> gravestone, determined Dooley's birthdate as April
> 4, 1878. These discrepancies can lead to confusion.
> Kind of like the game in which you whisper "duck"
> in someone's ear, followed by that

person whispering into someone else's ear and eventually the last person blurts out "giraffe." How can "giraffe" be attained from "duck"? Let's see:

1. The first person whispers "duck" to the second person.
2. The second person hears "buck" and whispers "buck" to the third person.
3. The third person hears "bug" and whispers "bug" to the fourth person.
4. The fourth person hears "grub" and whispers "grub" to the fifth person.
5. The fifth person hears "grab" and whispers "grab" to the sixth person.
6. The sixth person hears "graph" and whispers "graph" to the seventh person.
7. The seventh person hears "giraffe" and blurts out "giraffe."

The purpose of this verbose rendering is to demonstrate why going back to the source to verify information is important as opposed to being too reliant on interpretations. Interpretations, or misinterpretations, are influenced by many factors, such as translations, personal experiences, effectiveness of teaching, and attempts at application of principles in differing industries and environments, culture, and more. When a discrepancy exists, let's acknowledge it. If an assumption is not accurate, try to find the source of the discrepancy that may have contributed to the path that strayed. This thinking is in line with Lean practitioners' insistence to go to the *gemba,* or the place of the activity, to gain a deeper, less filtered understanding.

In 1902, Dooley started work with Westinghouse Electric & Manufacturing Company located in Pittsburgh. By 1911, he was the head of the organization's education department.[4]

Dooley provided advisory support from 1918 to 1919 around World War I to the US War Department as a director to the vocational committee on education and special training.[5] During the war, Congress deemed it necessary to increase production of materials while foreign commerce was interrupted.[6] Dooley developed relationships with government officials who in turn acknowledged the importance of accelerated production during wartime.

Dooley gained much experience in learning how to train others during urgent situations. By 1934, he had participated with the Federal Committee on Apprenticeships. In 1940, he was asked to organize and direct the National Defense Committee's TWI program.[7] Dooley was joined by three other experts in industrial training: Mike Kane from American Telephone and Telegraph, William Conover from US Steel, and Walter Dietz from Western Electric.[8]

LENS GRINDERS

The United States entered World War II in December 1941. The war ended less than four years later. Some of the manufacturing processes took more than four years to master. Citizens who previously made up the nation's workforce were now soldiers. The war effort demanded products quickly, so a new approach to training was needed.

In the summer of 1940, before the United States entered the war, one of Dooley's first challenges was to address a critical shortage of lens grinders. That particular craft took several years to master. Training needed to be reduced to only two months.[9]

Lens grinding required twenty different skills and a mastery of numerous process steps to do the job effectively. How did TWI address this problem? Have you ever heard anyone ask, "How do you eat an elephant?" The answer, "One bite at a time." In essence, that was what Dooley did. He simplified work into smaller bites. Individuals had to learn only part of the process, not the whole process. By analyzing the job, Dooley removed much of the learning that came from trial and error. What took years to learn before now only took days.[10]

UNDERWRITER'S KNOT

Dooley, who had an electrical engineering degree, enjoyed teaching people how to tie an electrician's knot, also known as the underwriter's knot—something Dooley could do quickly.[11] He showed a trainee and then had the trainee try; they typically failed miserably in the attempt. Dooley enjoyed seeing the trainee's eyes light up when able to tie the knot correctly after proper training.[12]

Underwriter's knot

Dooley developed a seven-step training process, which included:
1. The trainer showed the trainee how to do the task.
2. The trainer developed key points.
3. The trainer showed the trainee again.
4. The trainer had the trainee perform the simple parts.

5. The trainer helped the trainee perform the entire job.
6. The trainee performed the entire job while the
   trainer observed.
7. The trainee performed the job alone.[13]

The fundamental instructional theory that Dooley used was the concept of learning by doing, which focused on training and was fundamentally different from education.[14] It had been stated that education rounded out an individual for the good of society, but training helped to directly solve production problems.[15] The United States needed training at this critical point in history and not as much education. Learning by doing trained workers and directly addressed the production needs of the nation.

TRAINING WITHIN INDUSTRY

Training Within Industry had one simple focus: increase production.[16] Productivity, or doing more with less, was an objective of the training because the TWI team had to address the issue of limited resources and skills while still satisfying the burden of heavy demand for production.

It was impossible for Dooley and his team to personally train everyone in the country, so they focused on a scalable approach in which trainers trained other trainers. This had a multiplier effect in which trainers could train others without having to depend on large class sizes. Training classes were small, and these smaller class sizes made it possible for TWI to use the preferred concept of learning by doing.[17]

A critical success factor for achieving a multiplier effect was to train to a consistent standard. Trainers were trained to follow a specified outline. Since each trainer trained the same way and to the same outline, trainers were interchangeable. In the event a trainer left while giving a session, another trainer could walk into the room and pick up exactly where the previous one left off.[18]

There were four primary programs under TWI: Job Instruction Training (JIT), Job Methods Training (JMT), Job Relations Training (JRT), and Program Development. Each program used a training card that followed a four-step method on how to execute the respective program.

JOB INSTRUCTION TRAINING (JIT)

The JIT program focused on training trainers how to train workers to do their jobs correctly and conscientiously.[19, 20]

The JIT four-step card basically stated the following:

1. Prepare the worker.
2. Present the operation.
3. Try out performance.
4. Follow up.

Think of the card as a checklist or a reminder of the different steps. The list did not go into exhaustive literary prose of everything to do but rather served as a reminder. As Dr. Atul Gawande wrote in his book *The Checklist Manifesto*, checklists remind "us of the minimum necessary steps" and make them explicit.[21]

I took a senior leader to the TPS overview workshop in Georgetown, Kentucky, in 2010. One of the sessions was a demonstration of job instruction. He volunteered to be trained using the JIT method. He was stunned at the JIT effectiveness, and he immediately embraced Toyota's methods. He coordinated with Toyota to help him transform his production system.

Learning using JIT

JOB METHODS TRAINING (JMT)

Job Methods Training focused on pulling ideas for improving methods[22] and the way jobs were done[23] and was deployed by September 1942.[24] By October 1945, there were 377,213 certified supervisors in JMT.[25]

The JMT program required an assurance that workers with ideas took the time to physically write down their suggestions if the program were to have any chance of success.[26] Credit was given to the originator. Stolen ideas were believed to degrade morale and stop workers from being willing to give new ideas.[27]

This training taught supervisors to make regular incremental improvements and make the best use of resources available, such as personnel, machines, and materials. The program also taught them to use the scientific management principles to design work efficiently and not simply force a speedup in production.[28]

The JMT four-step card stated:

1. Break down the job.
2. Question every detail.
3. Develop the new method.
4. Apply the new method.

JOB RELATIONS TRAINING (JRT)

People problems, or issues, tended to occur in the workplace and they were considered production problems. After all, a person was hired to produce. A method was needed to provide a way for leaders to recognize such problems quickly and to systematically resolve those issues in a timely manner. Training Within Industry developed JRT.

The JRT program helped leaders address relationships between the supervisor and the worker.[29] It was released in 1943 and focused on treating work problems as production issues. A step-by-step method was created to provide a scientific approach to address people issues.

The JRT four-step card stated:

1. Get the facts.
2. Weigh and decide.
3. Take action instead of passing the problem to someone else.
4. Check results.

PROGRAM DEVELOPMENT

This tool helped organizations identify and determine the training programs needed for the workforce.[30]

The Program Development four-step card stated:

1. Spot a production problem.
2. Develop a specific plan.
3. Get plan into action.
4. Check results.

Training Within Industry could be pulled off the shelf and applied today if there was ever a need to increase production at a substantial rate. Of all the programs, it was JIT, with little to no modifications over the years, that stood as the single-biggest enabler to drive competencies and quickly improve production at greater speeds with superior quality.

WORLD WAR II

France fell on June 22, 1940, to Nazi Germany. The US government responded with the formation of TWI in August 1940 as an emergency service to improve production in an effort to support the United States and its allies.[31] Dooley, Mike Kane, William Conover, and Walter Dietz brought a wealth of knowledge and experience to the table; Dooley and Dietz each had thirty-eight years of industrial experience.

The TWI team addressed the growing needs of the country on a national scale, so the TWI staff grew to over 400 by 1944.[32]

The Great Depression left eight million people unemployed by the early 1940s. Many of the unemployed had never stepped foot in a factory or a shipyard.[33] The expected production needed for the war required unskilled workers to perform simpler tasks.[34] By 1942, 6,000 new workers, primarily those who were previously unemployed, were reporting to work daily, expanding production into extra shifts to include night shifts.[35] Training

Within Industry used its programs to accelerate the training of this growing workforce.

Supervisors were key stakeholders and players in the successful deployment of TWI in each organization, but they needed to be empowered. Thus, TWI emphasized to business leaders that supervisors required: (1) knowledge of the work performed, such as industry tools and technical skills, (2) knowledge of responsibilities, such as company rules and contracts, and (3) skills in instruction, improving methods, and leadership.[36]

A business leader's judgment of program success was a simple criterion: Did production increase?[37] That simple criterion did not necessarily mean that business leaders were going to immediately embrace TWI. Many business leaders often demonstrated trepidation because in their minds their production situation was different than other businesses. They questioned the applicability of TWI.

Ultimately, the underlying issue for any industry or business was that training methods were inadequate, and TWI filled that gap. It was determined that about 80 percent of the problems in production were attributed to poor training.[38] It was not good enough to put green workers side by side with experienced ones so the trainee could watch and learn.[39] Workers needed to learn by doing.

A new workforce

To create the TWI programs and make them workable, the TWI founders came up with four essential requirements:

1. Simplicity
2. Minimized presentation time
3. Built on the principle to learn by doing
4. Have a multiplier effect in which those trained could train others.[40]

To deploy TWI to millions of workers, TWI founders employed a multiplier effect with the goal of training the trainers. The bottom of each page of their manuals stated: "WORK FROM THIS OUTLINE—DON'T TRUST TO MEMORY." Deviation from the manual was not permitted.[41] A conference technique was adopted in which the trainers taught new trainers how to use the manual so they may train others.[42]

Job Instruction Training was trialed in seventy plants over six months.[43] Trainers taught leadmen who in turn taught the workers.[44] By October 1941, JIT was formally put into operation. The train-the-trainer concept achieved an impressive training rate of over ten million workers out of the sixty-four million that made up the country's workforce.[45]

The TWI program was effective almost immediately. In the year prior to the US entry into World War II (June 1940–June 1941), the nation saw the following production increases related to the TWI program:

- airplanes, 300 percent
- tanks, 600 percent
- gunpowder and ammunition, 1,000 percent

After the bombing of Pearl Harbor, the American war industries expanded from headcounts that numbered in the hundreds to headcounts in the thousands.[46] These workers consisted of women, high school boys, white-collar men, the handicapped, old-timers, and people from all walks of life. They needed a good training program with good supervisors.[47] The TWI program helped with both.

By September 1945, the six hundred clients monitored by TWI reported the following results:[48]

- Production increased by 86 percent.
- Training time reduced by 100 percent.
- Scrap reduced by 55 percent.
- Grievances reduced by 100 percent.

The TWI program issued 1,750,650 certificates during the war, most of which were for job instruction (1,005,170).[49]

After World War II, TWI as a government program was no longer needed, so it was disbanded on September 28, 1945[50]. Dooley and his team documented their work in the TWI Report, which was disseminated to the forty-eight states.[51] The TWI story was only beginning.

POSTWAR UNITED STATES

After World War II, the United States was the only country whose industry was still intact with no fear of competition. US companies produced to satisfy the demand in the United States and countries across the globe. With such a favorable position, these US companies may have argued, "Don't fix what's not broken." There was complacency building toward continuous improvement.

Automobile companies started to supply larger cars to the United States public. The demand for larger cars was the result of good road systems and cheap gas. There was very little demand for small cars. The US automobile companies sought other countries to import smaller vehicles for the smaller market. This market difference made the US vulnerable to the inflated oil prices of the future.

Overseas, especially in Japan, the situation was much different. There was no room for complacency. Japanese producers wanted to increase production quickly. Japan was ripe for TWI.

POSTWAR JAPAN

After the war, Japanese industry suffered, with production levels drastically less (only at 10 percent) than 1935–1937 production levels.[52] Production needed to improve. To stimulate the economy, General MacArthur's occupation, which had a dominant presence in postwar Japan, authorized the use of TWI to increase productivity on a national scale.[53]

The Economic and Scientific Section (ESS) under MacArthur's command, had oversight of the TWI program in Japan for seven years.[54] General MacArthur had several staff members who previously worked in the War Manpower Commission during World War II and were familiar with TWI.[55]

TWI Inc., led by Lowell Mellen, was the company selected to introduce TWI to Japan.[56] The company used the multiplier effect, just like in the United States during World War II, to create a comprehensive and long-lasting program.[57] Mellen brought three TWI instructors with him and contracted with the Japanese Labor of Ministry to deploy TWI's three programs: JIT, JMT, and JRT. Each was led by one of the instructors; a fourth was brought in who specialized in plant installations.[58] TWI Inc. used the press, the government, and military authorities to create publicity for them, which helped to provide credibility to the program and popularize TWI.[59]

Though TWI was brought to other war-torn countries after World War II, it was Japan that embraced the programs the most.[60] There were many cultural elements that may have helped the Japanese people embrace TWI. Japanese alliances, known as *keiretsu*, were common. These alliances tended to encourage organizational learning across traditional corporate boundaries to enhance a business's learning ability.[61] Also, junior managers were rarely promoted past senior managers no matter how well they performed, eliminating the fear of displacement.[62] Finally, eighteen-month to two-year job rotations in various functional areas, along with excellent education and high literacy rates, helped.[63] With TWI, Japan achieved a minimum of a 25 percent increase in production.[64]

The Japanese embraced the JIT and JMT programs, but the JRT program created some controversy because Japan felt JRT was a reactive way to address a problem that was already present versus preventing that problem from occurring.[65] Companies like Toyota took proactive steps to create cultures that prevented the need for a reactive program such as JRT.

TWI Inc. departed Japan leaving thirty-five certified institute conductors at the conclusion of its commitments. These conductors continued to use the multiplier effect to train trainers. By 1966, more than one million Japanese managers and supervisors were trained.[66]

Toyota was an early adopter of the TWI programs[67] when starting the program in 1951.[68] Toyota renamed their TWI program Toyota TWI.[69] In the 1950s, Toyota altered the JMT program to reflect the work initiated by Taiichi Ohno in creating the TPS.[70]

Worker ideas were valued. Supervisors at Toyota responded to ideas from workers within twenty-four hours. By 1986, 96 percent of suggestions given by workers were implemented.[71] The TWI program naturally influenced Toyota's standardization philosophy to ensure consistency and sustain ideas proven to work.[72]

ONGOING IMPORTANCE

If TWI's effectiveness in the United States was a fluke, then that fluke repeated itself in postwar Japan. Having been tested by two vastly different cultures with a high sense of urgency to produce at elevated levels with superior quality, TWI could one day again be pulled off the shelf at a national level if similar crises recurred. Somehow, Dooley and his team discovered a time-tested way to train people and invigorate production for industries in need.

Dooley's work with TWI did not go unnoticed. He received an honorary doctorate degree from Purdue University in 1944[73] and was inducted into the Academy of Human Resource Development Scholar Hall of Fame in 2000.[74]

SCIENTIFIC MANAGEMENT

The Japanese incorporated various aspects of scientific management into their programs such as quality circles, an initiative that engaged workers to regularly improve quality.[75] Specifically, scientific management ideas, including job simplification and motion studies,[76] were incorporated into the JMT program. The traditional scientific management approach was typically applied in a command-and-control manner, whereas TWI converted the approach to more of true management by science.[77] In other words, they, specifically Toyota, studied the work to determine the most effective way of performing tasks with consistent results. They then documented the work and trained all relevant parties to that standard.

OTHER POINTS

Postwar occupation by the United States determined there was a need for trucks to help rebuild Japan. Toyota helped to manufacture those trucks.[78] The market for trucks gave Toyota a business need, and TWI accelerated the training necessary for Japanese workers to deliver Toyota trucks consistently and effectively. Toyota credited TWI with developing elements of their TPS, especially in the areas of managing continuous improvement. [79]

Toyota in the 1980s began teaching Americans about the TPS. John Shook, an American manager who worked for Toyota and future owner of the TWI Network, Inc., admittedly expressed his frustration in trying to embrace the Japanese approach to training. He felt it "was too standardized, rigid, and rote in nature." Shook wrote of his amazement when his Japanese colleague, Isao Kato, showed him the origin of the Toyota training program by retrieving a well-worn TWI training manual. Toyota was simply teaching the Americans the same thing Americans had taught them.[80]

With any program, leadership is a critical ingredient. Poor or incompetent leaders could take a great program or concept and lead it to disastrous results. The TWI program had good leaders. As irony would have it, TWI was born during a time of great violence associated with a world at war, so how fitting that such a program was led by a very calm, never angry, leader by the name of Channing Rice Dooley.

CHAPTER 7

KIICHIRO TOYODA
(JUNE 11, 1894–MARCH 27, 1952)

"We shall learn production techniques from the American methods of mass production. But we will not copy it as is. We shall use our own research and creativity to develop a production method that suits our own country's situation."[1]
–KIICHIRO TOYODA

JUST-IN-TIME

The son of Sakichi Toyoda, Kiichiro grew up as a sickly and frail boy showing little ability to lead.[2] The opposite was true. Kiichiro founded the world-renowned automobile company whose name, Toyota, became synonymous for best in class. Kiichiro's leadership differed from his contemporaries because of his integrative view of business, his inventive approach to designing a production system, his hands-on approach to product development, and his view that people learn by doing, a belief shared by Channing Rice Dooley.

Kiichiro studied mechanical engineering in the 1920s at Tokyo Imperial University.[3] By 1928, Kiichiro was putting together the production system to manufacture automobiles in Japan.[4] In 1929,

Kiichiro traveled abroad to negotiate patent rights to sell the Platt Brothers rights to produce Toyoda's automatic loom. Kiichiro also visited America because it was the premier source of automobile manufacturing in the world. While there, Kiichiro studied Ford's production systems. Upon Kiichiro's return, he worked on the foundations for the Toyota Motor Company.[5]

The Ford plants manufactured automobiles on a large scale with a considerable amount of money tied up in inventory throughout the manufacturing process. Kiichiro did not have the luxury to use the same kind of production system in Japan. Though Ford embarked on initiatives such as waste reduction, it was still too excessive for Toyota because of the comparatively large amount of cash tied up in the production process. Additionally, Ford's efficiency was attained partly by the American market allowance for few model variations that could be produced in great volumes. Kiichiro needed to adopt a way that produced a greater variety of automobiles in which each variation had lower volumes demanded by the Japanese market.[6]

While visiting America, Kiichiro observed how grocery supermarkets maintained a regular supply of various products at relatively small volumes to maximize the use of selling space. Kiichiro noted that a product was restocked when a customer pulled it off the shelf for purchase.[7] The empty slot triggered a need for replenishment. This was the kind of thinking Kiichiro needed to create his production system, one that minimized inventory by producing only what would sell to keep cash flow a constant.

SEQUENTIAL STEPS

In the late 1930s, Kiichiro organized assembly production stations into sequential order and positioned them in proximity to each other. It seemed like it would be common sense to put step one before step two and in the same proximity, but some organizations reasoned the key to efficiency involved colocating all like processes with little regard to the flow of production throughout the entire process.

If efficiency was measured by how departments operated in silos without any regard to any interdependencies with other production stations, then an argument could be made to support that notion. If efficiency was measured by how well all production stations worked together to facilitate the flow of products, then thinking in silos would degrade efficiency because of excess inventory between each station. Excess inventory correlated to longer lead times and increased risks for the presence of quality issues.

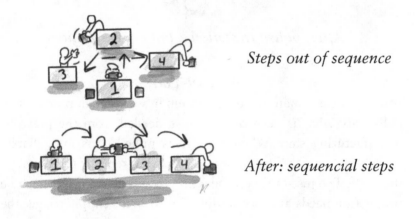

Steps out of sequence

After: sequencial steps

PRODUCE ONLY WHAT IS NEEDED

Kiichiro designed a production plan to only manufacture products at each station when needed.[8] Thus, products were produced just-in-time. That simple philosophy was the basis for Kiichiro's production system, and it served as the founding principle of what would become known as the "*kanban* system."[9] Toyota Motor Company credited Kiichiro as coining the term "just-in-time."[10]

Just-in-time manufacturing was adopted based on the assumption that inventory was not cash. Money was money.

Before: pushed to station 2 (excess inventory)

After: pulled to station 2 (minimal inventory)

JUST-IN-TIME

Just-in-time is a manufacturing system in which each process step pulls only what is needed when it is needed from the previous manufacturing step. When an item is pulled, it is immediately replenished, much like how supermarkets replace products on the shelf. The just-in-time production schedules do not determine production needs at every station. Pulled items simply trigger the need for production upstream, helping to minimize inventory levels. The just-in-time concept is not dependent on a rigorous centralized scheduling system using complicated formulas.[11]

> Several years ago, I was a production supervisor at a manufacturing facility responsible for finishing (coloring) operations of casted parts. The materials management team, which owned production scheduling, had a practice of scheduling large batches of casted products to ensure there were plenty to satisfy any variation to production needs. This created a massive inventory of

unfinished casted products. Inventory was parked everywhere. New shelves were constantly built to store this ever-increasing inventory.

Despite this large inventory, we often did not have what we needed to satisfy demand. To keep my department running, the scheduling team instructed us to produce products with no demand. We were required to be busy, but not necessarily productive. Products with demand often required expediting.

This inefficient scheduling approach caused us to work harder. We staffed four shifts with over 140 people covering all hours of the day, seven days a week. It felt like we were chickens running with our heads cut off.

The growing inventory became more problematic because shelves were being damaged by forklifts. These shelves were more than two stories high holding several tons of products, and I feared one day they would all come tumbling down.

Eventually, through the wisdom of new plant leadership, the materials manager was replaced with someone who focused on producing only to demand. We were able to get the casted parts we needed when we needed them. The casting department worked closely with my department to ensure casts met quality standards. Inventory in storage was reduced, and shelves were able to safely store what remained.

Within six months, we reduced operations to five days a week, and we required less than ninety employees. We reduced our workforce through normal attrition and transfers with no layoffs. Cash flow improved because money was not tied up in inventory. Costs were down because we were not investing in producing product with no demand. Safety improved because of less inventory and clutter. Quality improved because departments had time to work together to discuss quality requirements.

The employees became much more engaged in mentoring each other, continuously improving processes and improving the facility's safety program. New leadership and training positions were created, which further empowered and motivated the workforce. Several awards were given to multiple departments for operational excellence. Most notable was the success the entire plant achieved working as part of the Occupational Safety and Health Administration's (OSHA) Voluntary Protection Program (VPP). The facility earned the OSHA VPP Star status, the highest recognition for an exemplary safety culture. This rudimentary application of just-in-time transformed our plant.

In 1934, Kiichiro manufactured the first Toyota engine. By 1935, the first automobile prototype was produced.[12] Soon, Kiichiro Toyoda became Toyota Motor Company's first president.[13] It was Kiichiro's goal to produce and sell cars at reasonable prices to the public. As the president, he understood the importance of having good materials for use in production and the role of sales in the automobile industry.[14]

In 1936, Japan put the Enterprise Law into effect, giving automobile companies such as Toyota the ability to compete freely without worrying about the dominant American automobile companies.[15] Despite the help from the government, Kiichiro did not let complacency set in. He maintained his focus on producing affordable, good quality cars.[16]

Japan's defeat in World War II could have brought automobile production to a halt for Toyota, but it did not. During the postwar-US occupation, the Americans determined that trucks were critical to rebuild Japan.[17] Toyota, therefore, began producing trucks.

Kiichiro expected his employees to get their hands dirty if necessary. During a walkthrough of one of his facilities, Kiichiro saw a worker who could not figure out how to get his machine to work, so Kiichiro put his hands into the machine's oil pan and pulled out two handfuls of sludge. Kiichiro said: "How can you expect to do your job without getting your hands dirty!"[18]

Kiichiro led the start-up of Toyota's automobile entity and carried it through the turbulent times of World War II. Toyota endured the challenges of postwar Japan during a period when the country was trying to rebuild itself. No matter how good of a production system Toyota had, it was not perfect. Additionally, the economic, political, and environmental influences would eventually take its toll on Toyota's financial health.

By the end of the 1940s, Japan's economic inflation created havoc on the economy. Kiichiro, to avoid bankruptcy, had to cut costs, which included pay reductions for his employees. Toyota's production system cash was getting tied up in inventory because of low demand despite their efforts to minimize inventory.[19] Kiichiro had a no-layoff policy, but that policy started to cause additional financial distress. Eventually, Kiichiro had to break his promise.

In May 1950, Kiichiro was forced to lay off approximately one-third of his employees (2,146). The layoff was followed by a two-month long strike.[20] Kiichiro accepted the responsibility for the situation, so he resigned from his post as president. Kiichiro's resignation helped to ease tensions between Toyota and the employees.[21] It seemed Kiichiro's resignation was a very low point for the struggling automobile company.

The economic and political situation soon changed with the breakout of the Korean War about one month after Kiichiro's resignation. Toyota Motor Company. began filling production orders to support the war effort. Kiichiro died in 1952.

Under Kiichiro's leadership, Toyota incorporated the concepts of interchangeable parts, scientific management, flow, pull, *jidoka*, and learning by doing. Kiichiro's work was expanded upon by others, and new innovations improved the just-in-time system. Eventually, Toyota's business model could endure both economic and political strife. Toyota made Kiichiro's just-in-time model and Sakichi's *jidoka* concept the pillars of its production system, which was both fluid and responsive to demand. Those pillars were built on a culture that respected, engaged, and empowered its workforce.

CHAPTER 8

SHIGEO SHINGO
(JANUARY 8, 1909–NOVEMBER 14, 1990)

"Frequently, I am rebuffed by people who say they are too busy and have no time for such activities. I make it a point to respond by telling people, look, you'll stop being busy either when you die or when the company goes bankrupt."[1]
—SHIGEO SHINGO

SINGLE-MINUTE EXCHANGE OF DIES (SMED)

Shigeo Shingo was born in Saga City, Japan.[2] As a student, Shingo became familiar with the work of Frederick Taylor through the translated publication *The Principles of Scientific Management*. Taylor's ideas influenced Shingo's thinking considerably. Later in his life, Shingo became a consultant. He was probably the world's first Lean consultant. In that role, Shingo used Frederick Taylor's ideas to instill quality at Toyota.[3]

In 1930, Shingo earned his industrial engineering degree from Yamanashi Technical College. During World War II, he served as a production manager at the Amano manufacturing plant in Yokohama. While at the Yokohama plant, he was able to double productivity.[4] Much of his time between the 1940s and the 1950s was spent working on ideas to improve manufacturing efficiency.

Shingo introduced many ideas for manufacturing systems, but probably the most profound was Single-Minute Exchange of Dies (SMED), which was synonymous with quick changeover. In manufacturing, changeover involved the activities necessary to change a production line from one product to the next. For presses, it could involve changing out dies (hence, SMED). In health care, operating room changeover could include all activities necessary to remove one patient, clean the room, and prepare for the next patient. In all cases, changeover was downtime, so the goal of changeover improvements was to reduce downtime.[5] For SMED, "Single-Minute" referred to setting a goal to reduce changeover downtime to a single digit, such as nine minutes or less.[6]

Shingo developed the techniques for SMED in 1950 while conducting improvement work at Toyo Kogyo (Mazda) in an effort to eliminate bottlenecks from body molding presses.[7] During one occasion, Shingo observed an operator perform a changeover. The operator discovered a missing bolt and proceeded to conduct a lengthy search while the machine stood still. It was at this time Shingo realized that activities could be classified as either internal or external.[8]

Internal activities, or tasks, could only be performed while a machine was shut down, such as replacing the die. External activities could be performed while the machine was still operating, such as gathering the tools necessary to perform the changeover.[9] Another example of an external task included preheating the die and setting it near the press. When it was time to shut down the machine, all that was needed to complete the changeover was to perform the internal tasks, which were primarily focused on switching out the dies.[10] The science of classifying activities as external or internal tasks lent itself to the creation of a three-step methodology to reduce changeover time:

- STEP 1 involved the classification and separation of internal and external setup tasks. This step alone could reduce the downtime during changeover by as much as 50 percent.[11]
- STEP 2 involved converting internal setup tasks to external setup tasks.[12] This process could include a physical modification to the machine in question.[13]
- STEP 3 involved streamlining the setup tasks through techniques such as performing tasks in parallel and the use of clamps. Other activities could include the elimination of redundant tasks.[14]

In the 1950s, Taiichi Ohno invited Shingo to Toyota. One of the areas addressed was the reduction of setup time for equipment, which typically took hours. Toyota wanted to reduce the setup time to a matter of minutes.[15] There were only minimal improvements made at that time, but the urgency still existed. Toyota avoided changing dies on presses back in the 1940s because the changeover process would take several hours to complete—time in which the machine was not producing product.[16]

By the 1960s, Toyota was in desperate need to reduce the setup time, so in 1969 Shingo returned to consult for Toyota to tackle the changeover problem. Shingo worked with Toyota for six months making improvements[17] and reducing the setup time for a 1,000-ton press from four hours to a staggering three minutes.[18] The three-minute changeover itself was not achieved while Shingo was present. Drastic improvements to the setup time had already been made when Shingo had to leave the facility for a short period. Shingo was amazed at what Toyota had accomplished on its own.[19] It was Taiichi Ohno, the man who invited Shingo to Toyota, who had pushed the three-minute target. In Shingo's opinion, Ohno's insistence on reducing the changeover time to three minutes marked the birth of SMED.[20]

Shingo consulted with other companies, and it was common that SMED was met with resistance from workers. Shingo noted that leadership commitment was necessary to overcome obstacles like the complaint that SMED would cost too much.[21] Shingo often addressed challenges and resistance by performing trial runs. In one example, a trial run was performed using Shingo's techniques, and the changeover time was reduced from eighty minutes to less than eight minutes, a feat that quickly changed workers' attitudes to adopt SMED.[22]

In time, Shingo put together a demonstration for clients and prospects that consistently reduced setup times from over ninety minutes to less than eight minutes.[23] Sometimes, changeover time could be reduced to less than one minute. If that was achieved, then the term "One-Touch Setup" could be applied.[24] Ultimately, the effort was to create an instantaneous changeover to augment one-piece, or continuous, flow.

SMED played an important role in reducing in-process inventory and facilitating continuous flow of products because changeovers occurred more frequently. The reduction in large inventories significantly contributed to shorter lead times. Production lead times decreased from several days to only a matter of hours.[25]

One approach many organizations in health care use to facilitate quick changeovers in the OR is what they refer to as a "pit crew." Many people are engaged to complete the changeover quickly, but there are often inefficiencies.

Efficient changeovers involve more than just throwing people at a task. A standardized process with sequenced tasks and assigned responsibilities are also needed. A good example is a race-car pit crew. Race car pit crews have various internal tasks, or tasks performed while the car is present, assigned to varying roles. Someone cleans the windshield, another raises the vehicle, a team replaces the tires, someone refuels, somebody hydrates the driver and so on.

Before the changeover begins, the pit crew plans what is needed for the changeover, and they prep while the car is still racing. For example, someone may receive a weather report and decide to replace the tires with something more well suited for projected road conditions. The pit crew will prepare the tires accordingly. These external activities are performed before the pit stop. The same thinking, to separate internal and external activities, helps other industries maximize efforts and reduce downtime.

In health care, many surgeons want to minimize the time of OR changeover. This helps them reduce the wait time between surgeries so they can operate on more patients in a shorter period. I am often asked, "What is a good changeover time for an OR?" My answer is usually, "What does it need to be?" This may sound inconsistent with SMED because SMED focuses on changing over as fast as possible, but SMED times need to be consistent, repeatable, and reliable so production can be scheduled accurately.

Several hospitals I visit have embarked on quick changeovers in the past, but the improvements did not last. Often, the reason for this was that the OR remained open without a patient for lengthy periods after the changeover was complete. The urgency to changeover the OR was diminished.

I have seen very fast OR changeovers, but many were inundated with safety risks, redundancies, and incomplete work. I saw a changeover from the time a patient left the room to the entry of the next patient take only seven minutes, but the floors were not completely cleaned around the surgical bed, trash was contaminated with blood, and not all the relevant tables were wiped down.

For high volume, emergent surgical cases, a quick changeover is critical. I am not convinced the same exists for the vast number of elective procedures such as knee replacements, bariatric procedures, and many more. In such cases, I recommend considering two things. Determine how long the changeover *needs* to be. Determine what meets the patient's *expectation(s)*? Other guiding suggestions are also given.

In some cases, especially when the coordination of activities between numerous departments and caregivers exist, it is more desirable to have a changeover completed to a standard time and ensure the next case starts on time and as planned.

What does the changeover time need to be? Consider the following calculation based on a simple idea: time available versus time required. If there are eight hours available in a day for one OR, and eight cases are scheduled, then the OR needs to have one patient completed each hour. If each case lasts forty-five minutes, then the turnover only needs to be fifteen minutes (sixty minutes available minus forty-five minutes procedure time equal fifteen minutes).*

Standard OR changeovers facilitate on-time starts throughout the day for elective procedures. Many hospitals focus primarily on starting their first case on time with the reasoning that cases will follow and start on time also, but that thinking disregards the importance of standard changeovers. For elective surgeries, designed changeovers based on what can be achieved consistently contributes to regularity in the achievement of the surgical schedule as planned. If you can get the changeover time down to less than ten minutes repeatedly, then consider ten minutes. In such cases, it may be more apt to have a standard changeover versus a quick changeover.

There are many different variables when planning quick changeovers and determining such goals. In some cases, well-coordinated pit crews are immensely valuable; in other cases, a more steady and rhythmic approach is needed. Of course, there are clinical, economic, and operational factors to consider, but beginning with a patient-centered approach that considers the needs of the patient may help reach a decision.

*This is an oversimplification. Other factors to consider are break/lunch time, first and last case setup, end of day tear down/terminal clean, type of procedures, and so on.

OTHER CONTRIBUTIONS

Shingo had several other significant contributions to Lean. He pioneered the concepts of mistake proofing, also called *poka-yoke*, to prevent a mistake from occurring. Engineering two different connection ports that prevented someone from connecting a line to the incorrect port was an example of mistake proofing. Another concept Shingo introduced was source inspection. Inspecting at the source prevented passing defects to a process step downstream.

Around the 1950s, Japan was introduced to statistical methods to improve quality. Initially, Shingo believed in the value of statistics in production but eventually found himself at odds with the over-emphasis on statistical methods to enhanced quality. Shingo felt this kind of thinking created an acceptance of expected defects versus defect prevention. Concepts such as statistical process controls were based on the premise that 100 percent inspection was impossible. Shingo believed 100 percent was possible through such pioneering concepts as *poka-yoke* and source inspection.[26] This kind of thinking became an extension of Sakichi Toyoda's *jidoka* concept.

CHAPTER 9

W. EDWARDS DEMING
(OCTOBER 14, 1900–DECEMBER 20, 1993)

"You don't get ahead by making a product
and then separating the good from the bad."[1]
–W. EDWARDS DEMING

PLAN-DO-CHECK-ACT (PDCA)

Japan honored W. Edwards Deming in two very distinctive ways. First, Japan established the Deming Prize in 1951 for an individual who had an accomplishment in statistical theory and for companies who achieved success in statistical applications. The Deming Prize was immediately considered a prestigious award in Japan and highly sought after.[2] Japan also honored Deming with the highly prestigious Second Order of the Sacred Treasure medal, crediting Deming's work for the rebirth of Japanese industries. Who was this man to whom so much honor was given?

W. Edwards Deming specialized in the use of statistical methods. He grew up in Wyoming and was named after both his parents: William from his father and Edwards from his mother (it was her maiden name).[3] Deming earned his PhD in physics from Yale.[4]

Deming was an expert on statistical sampling. He applied statistical sampling while working for both the US Department of Agriculture and the Census Bureau.[5] While at the Department of Agriculture, Deming was introduced to Walter Shewhart. Deming traveled regularly to New York to meet and study with Walter Shewhart, and ultimately, theories by Shewhart served as the basis for Deming's work.[6]

> Deming proposed a new approach to taking the 1940 census. Instead of polling every individual in the United States, Deming proposed a sampling program. He made the case that sampling enabled quicker calculations than the polling for a greater volume of information. As a result, Deming was asked by the secretary of commerce to take charge, for which he accepted.[7]

Walter Shewhart worked at Western Electric, where quality control problems were being addressed as far back as the 1920s. In Deming's book *Out of the Crisis*, he warned of the risk of businesses seeking the lowest bidder, resulting in lower quality and higher costs.[8] Deming got the idea for understanding the true cost of poor quality from Shewhart's 1931 publication, *Economic Control of Quality of Manufactured Product*.

Shewhart took work samples and performed statistical analyses to identify performance variations.[9] Shewhart created techniques to bring processes into statistical control.[10] A control chart was created to visualize the random variation of any worker's task and classify the results as either being acceptable or unacceptable. Results between the upper and lower specification limits were acceptable. Unacceptable results needed to be studied to identify the causes.[11]

Control charts made their way into the American industry during World War II with the help of Deming. He taught both engineers and technicians on statistical sampling techniques to aid in war-material production.[12]

After World War II, the United States was one of the few countries not drastically touched by the war. There was a high demand for American products without the worry of competition.[13] Production took priority over quality. Quality was minimized to inspection at the end of the line. By 1949, according to Deming, control charts had faded from use.[14] This caused Deming some frustration with American industries' commitment to quality. His concerns were usually ignored.

Deming, recruited by the Supreme Commander for Allied Powers (SCAP), traveled to Japan in 1947 to prepare for the 1951 Japanese census.[15] Japan showed little evidence of any postwar recovery at that time. Food was scarce, and children were suffering. Viewing the poor circumstances of the Japanese people left a lasting impression on Deming.[16]

The industrial situation in Japan seriously lacked quality. "Made in Japan" was synonymous with junk. Japan needed to import food, but its industrial production was valueless as a trade for food.[17] Japan could not feed its people based on what they grew themselves.

The Union of Japanese Scientists and Engineers (JUSE) invited Deming back to Japan in 1950 to give lectures on statistical quality control.[18] Kenichi Koyanagi, the managing director of JUSE, along with several colleagues, were already familiar with Shewhart's written works and lectures[19] because Bell Telephone Laboratories sent books on statistical quality control to JUSE, including Shewhart's 1931 book.[20]

Beginning with his arrival in Tokyo on June 16, 1950, Deming spent the next thirty years traveling regularly to Japan to give seminars and lectures.[21] Throughout his time there, Deming immersed himself in the Japanese culture.[22] Conditions had improved in Japan since his 1947 trip. People were clothed and better fed. Much to Deming's joy, they looked happy.[23]

Deming, during his lectures, helped to redefine "the customer," stating there were both external and internal customers. In a production line, the next process was the internal customer to the preceding process step.[24] Toyota embraced this thinking to help facilitate part of its just-in-time production system.

Though the Japanese people embraced statistical techniques, Deming worried their conviction would run dry, as it did in America, if Deming could not get the people in charge to care. Ichiro Ishikawa, the JUSE president at that time, set up a dinner with Deming and several of the country's chief executives. It was at this time that Deming urged them not to accept poor quality from vendors because it would affect the quality of the products the Japanese produced. Deming reasoned with these Japanese leaders that if Japan could improve its quality, then Japanese industry could capture market share in the world within five years; Japan achieved that goal in four years.

Deming used Chicago, England, and Switzerland as examples of places that didn't produce all their food but rather exported products as trade for imported food.[25] This exchange of value required quality products. Very quickly, charts and checklists could be seen in factories throughout Japan, and results were impressive. Rework was reduced to 10 percent of its original levels in one corporation; in another, production increased three fold. By mid-1951, products coming out of Japan started to be noted as being of good quality.[26] Within ten years, Japan had trained nearly 20,000 engineers in rudimentary statistical methods.[27]

Japan and Deming developed a mutual love and respect for each other. On the other hand, US industry was not very receptive to Deming's statistical methods. In time, he was able to make some headway with some US companies, such as Nashua Corporation.

Clare Crawford-Mason, a television producer, decided to film a documentary tentatively titled *Whatever Happened to Good Old Yankee Ingenuity?*, in which she focused on the Japanese economic threat to the United States.[28] When she reached out to Deming, he impressed her, and she decided to include him in the broadcast.

On June 24, 1980, during NBC's television program *If Japan Can, Why Can't We?*, Deming gained quick fame in the United States, primarily for his work with Nashua. Nashua's president at that time, William E. Conway, stated that under Deming's guidance, the company had saved millions of dollars and increased its productivity.[29]

Years before the broadcast, Japan had already gained much attention for improved quality and business practices when the oil crises of the 1970s caused unprecedented financial stress on US automobile manufacturing. There was a clear and distinguishable difference between the Japanese and American management systems.

Deming's attitude for US management in the 1980 NBC broadcast served as a blunt look at the American management system, and it was not a pretty sight. Deming mentioned, "It would be a mistake to export American management to a friendly country"—a highly differing point of view than the World War II and the immediate post–World War II eras.[30] Deming pointed out that 95 percent of errors were due to systems, not people. He also pointed out that 85 percent to 94 percent of issues faced by organizations were caused by management. Deming's harshness on managers stimulated an interest in quality management in the United States.[31]

Deming emphasized systems, not people, in his berating of US management. He developed fourteen points on this subject and identified seven deadly diseases that he believed caused the decline in US industry. Some of these diseases included lack of constancy of purpose toward improving products and services, an emphasis on short-term profits, job hopping by managers, and managing by numbers without considering unknown or unknowable figures.[32]

The United States listened, and Deming became extremely busy working with US companies to build quality into their organizations. He had clients that stretched from America, such as the Ford Motor Company and General Motors, to South Africa and from New Zealand to England.[33]

PLAN-DO-CHECK-ACT (PDCA)

Deming's most notable Lean contribution is PDCA in the role of problem solving. Also called the Deming Cycle or the Shewhart Cycle, PDCA is a way to improve upon systems.[34] The PDCA steps include planning, doing, checking/studying the results, and acting based on results and then planning for the next iteration.[35]

Much of PDCA involves breaking up processes to transform them into better or more effective processes. Described as the Juran Trilogy, named after Joseph Juran, the process moves systematically from a controlled state, to an out-of-control state, then to an improved state.[36]

The PDCA cyle is simply for improvement.[37] It is synonymous with PDSA (Plan-Do-Study-Act) with minor differences, and it is an indefinitely continuous cycle progressing with unending improvements.[38] Use of PDCA is pragmatic, and it guides ongoing organizational learning and growth, evidenced in the Toyota organization by how the company employed solutions, or countermeasures, and how it checked to make sure the solutions worked.[39]

At one point, Toyota noted that inventory needed to be reduced because of the inherent risk of quality issues as well as inventory's contribution to longer lead times. Toyota strategically reduced inventory levels to facilitate more flow and reveal new problems.[40] Toyota used PDCA to systematically change the inventory levels and stabilize production process at the new levels.

Initially, the Shewhart Cycle was a three-step approach: Specification (Plan), Production (Do), and Inspection (Check).[41] Deming introduced the more familiar four-step cycle. Near the end of Deming's life, he changed the "Check" in PDCA to "Study" (Plan-Do-Study-Act), something that he felt stayed true to Shewhart's original intent. He published the PDSA cycle in his 1993 book *The New Economics for Industry, Government, Education*.

They both followed the scientific method for problem solving and continuous improvement. This scientific method did two things: provided logic others could follow and helped to connect the dots between cause and effect so solutions/countermeasures directly addressed the causes and drove the desired effect(s).

Plan   Do   Check   Act

From the very start in 1950, Deming had the Shewhart Cycle on a blackboard for each of his lectures. He did not necessarily call them out as PDCA at first. His four steps at that time were:

1. Study the process, determine a change to improve, organize the team, determine data necessary, determine tests needed, and ensure a plan is put together (do not proceed without the plan).
2. Perform the tests or make changes (on a small scale preferably).
3. Observe the effects.
4. Evaluate lessons learned and, if necessary, repeat the tests. Pay attention for potential side effects.[42]

Those trained by Deming were the ones who optimized the cycle with the following four steps by 1951.

1. Plan: define a problem and hypothesize potential causes and solutions.
2. Do: implement the solution.
3. Check: evaluate results.
4. Act: if unsatisfactory, return to the planning step; if satisfactory, standardize the solution.[43]

At some point, it was decided to tell the entire problem solving/continuous improvement story on a single sheet of paper. The paper selected was an A3 size (11.7 inches by 16.5 inches). It was the largest sheet of paper that could be faxed at the time. The A3 template provided a discipline to follow each of the PDCA steps and prevent some of the most overlooked steps in problem solving: checking and acting.[44]

During my first visit to Toyota in Georgetown, Kentucky, one of the instructors was new to teaching TPS. The instructor was asked a question by one of the students, but she claimed she was not sure of the answer, so she looked to the lead instructor. The lead instructor encouraged her to answer to the best of her ability. She did, and the lead instructor, with a smile, said that she answered the question correctly. What humility!

After the class, with the same kind of humility I witnessed, I approached the lead instructor and asked her if she knew anything about PDCA (dumb question). Her voiced pitched with excitement and she invited me to sit down so she could explain it to me. It was obvious. It was simple. It was common sense. When asking the question, I thought I'd end up walking away with my head down. Instead, she made me feel empowered. The lesson: humility and a willingness to learn can go a long way to help teachers teach us.

STATISTICAL PROCESS CONTROL (SPC)

At the heart of much of Deming's teachings was the use of statistical process control (SPC), which was a quality improvement methodology that graphically displayed data in consecutive order so the process could be monitored and controlled.[45] The data could be used to improve processes.

Run charts and control charts are two common forms of SPC. Run charts visualize trended data to show variation. Control charts, on the other hand, have upper and lower control limits that team members can monitor to ensure the process is performing within specified parameters.[46] If data is plotted outside the control limits, then a special cause is responsible for the variation.[47] The team then investigates and addresses the special cause in an effort to bring the process back into control.

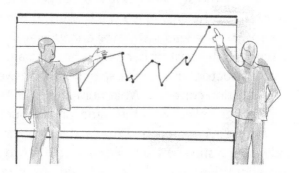

Address points outside the limits

Before Deming, Japanese production systems were no different than those in America in that they favored keeping production lines moving versus stopping the line to address a quality issue.[48] Toyota, among other Japanese companies, embraced Deming's teachings in the objective use of statistics to drive quality.

Taiichi Ohno and Shigeo Shingo learned from Deming and Juran a combination of techniques and methods that would work for TPS.[49] Toyota, with a clear understanding of the intent to build quality into a production system, created a culture that encouraged stopping the line to contain and fix quality issues immediately upon being discovered.

DEMING'S IMPACT ON TOYOTA

Toyota innovators consistently tested concepts and evaluated techniques in an effort to build its production system. However, it did not fully embrace all of Deming's teachings. The PDCA cycle meshed well with TPS; statistical methods did not appear to take as strong of a hold.

Toyota used PDCA as the step-by-step discipline to continuous improvement.[50] Through its methodical approach, PDCA prevented repetition of mistakes.[51] According to many experts, PDCA and standardization were core to both Lean management and the TPS. [52]

Toyota defined customers as being both internal and external based on Deming's ideas.[53] In this way, the company was able to communicate the quality requirements upstream from downstream processes and the ultimate customer. The workers in the upstream process steps committed to not dispatching a product to the next process step, also known as the internal customer, unless it met all quality requirements. Defects were stopped at the source.

An interesting note: Nissan won the Deming Prize in 1960, so Toyota, which was stung by this news, effectively intermixed and aligned both quality and continuous flow throughout the organization and won the Deming Prize in 1965.[54]

Toyota displayed the attributes of Deming's constancy of purpose. It did not focus on short-term profits but rather a slow and steady movement forward with little change in direction or strategy shifts and regular profits year after year.[55]

VARIED VIEWPOINTS

Deming's teachings did not achieve 100 percent consensus. With different cultures, different types of industry, different situations, and different levels of maturity in Lean systems, it was only reasonable that varying viewpoints would develop. The underlying theme that did persist was the objective approach to quality control versus the subjective approach, which could be riddled with fallacious reasonings.

There were a couple of varying viewpoints from others who had a respectable impact in driving quality in industry.

JOSEPH JURAN

Juran (December 24, 1904–February 28, 2008) worked at Western Electric and, like Deming, was also influenced by Walter Shewhart on the ideas of statistical quality control. Juran did not necessarily believe that quality control was only a matter of statistics.[56] He traveled to Japan in 1954 and approached the subject from the perspective that quality control was a job for everyone.

Juran used an effective teaching approach that helped managers and workers who struggled with Deming's views on statistical quality control. He also introduced the idea of focusing on the vital few that would have the biggest impact. One example was the 80/20 rule, which stated that 80 percent of the problems were caused by 20 percent of the operations.[57] Once the vital few were addressed, most problems were solved.

SHIGEO SHINGO

Shingo believed he had to overcome the overemphasis on statistics that Deming fostered. He argued the goal of quality was to prevent defects, whereas statistical methods were based on the expectation of defects.[58] Shingo was uncomfortable with SPC because SPC inferred that 100 percent inspection was impossible. Shingo proved that 100 percent inspection was possible at low costs using two techniques: mistake proofing (*poka-yoke*) and source inspection.[59]

While working as a production supervisor, I had a quality problem with the coloring of some of our products. It was sporadically too light. This had been a problem for approximately fifteen years, and I wanted to nip this one in the bud. I was tired of dealing with it.

With the support of my manager, we aggressively attacked the problem. We were still investigating the problem when our scrap dollars drastically dropped. An engineer in the quality department told us not to waste our efforts on this problem anymore. I objected. Scrap was down because the production volume for that color was low. If we did not fix it, the problem would recur.

One of our production steps, color buffing, was a manual process that directly contributed to the coloring of the product. I decided to investigate the differences in quality by each operator. Statistics helped me isolate one operator as having superior quality.

My team leader and I brought this high performer into my office to show him the statistics and asked if he had any insight as to why his results were far superior to the other seventeen operators. He had nothing to offer but was pleased to know he was performing well. I asked him to show us how he scratched the parts with the buffing machine.

We walked to the shop floor, approached one of the machines, and turned it on to the speed as indicated by the job instructions. The operator said the machine spun the buffing wheel too fast. All the machines were set up the same.

I asked, "What do you mean too fast?"

He said, "The wheel speeds don't usually match the readouts on the machine."

My shift lead retrieved a device that could measure the RPM (rounds per minute) of the buffing machine wheels. Sure enough, the wheels were not calibrated. They were running faster than what the RPMs showed on the display. In fifteen years, nobody had ever calibrated these machines.

Why is this significant? The statistical tools helped significantly, but here was an opportunity to build in quality. As Juran pointed out, fixing the "vital few" solves many other problems. In our case, we saved more than a third of a million dollars in scrap each year. Like Deming suggested, statistical process controls can be used to predict how long the machines can run before they need to be calibrated again. Ultimately, it was regular calibration and 100 percent inspection by operators to a standard that got us to our goal, as Shingo suggested.

Despite these varying perspectives on Deming's views of statistics and quality, all agreed quality was critical to success. On the NBC special *If Japan Can, Why Can't We?* Deming stated:

"Inspection does not build quality. The quality is already made before you inspect it. It is far better to make it right in the first place. Statistical methods help you make it right in the first place, so you don't need to test it. You don't get ahead by making a product and then separating the good from the bad."

Could we produce to the same level of quality as Japan? Absolutely!

CHAPTER 10

TAIICHI OHNO
(FEBRUARY 29, 1912–MAY 28, 1990)

"The production line that never stops
is either excellent or terrible."[1]
–TAIICHI OHNO

TOYOTA PRODUCTION SYSTEM (TPS)

The TPS owes its existence largely in part to one man, Taiichi Ohno. Ohno was the central figure who brought numerous disciplines and concepts together to form the (TPS). His personal innovations were groundbreaking, TPS had the biggest influence on what became known as Lean.

Ohno was unique compared to other Toyota figures. First, he was born in China, not Japan, in 1912.[2] Second, his leadership style was less tempered. Like many change leaders today, Ohno dealt with frustrations related to ongoing change and an unwavering commitment to his goals. These frustrations may have amplified his leadership temperament.

Like Shigeo Shingo, Ohno was inspired by the writings of Frederick Taylor, especially around instilling quality control.[3] He studied the works of Henry Ford. Along with other Toyota representatives, Ohno visited the automobile manufacturing plants in the United States.

To create a "lean" production system, Ohno needed a very skilled and motivated workforce. Workers needed to anticipate and correct problems before they occurred, thus reducing the risk of shutting down the production line. Training Within Industry programs were critical in helping Ohno achieve this by training the workforce and getting the workforce to participate in *kaizen* activities.[4]

Ohno valued data, even stating its importance in manufacturing, but was cautious because data did not represent real-time information and was merely an indicator. He valued facts, and that required going to the scene of activities to attain those facts.[5]

Almost anyone who ever attempted to introduce Lean into an organization would probably say it was very difficult, and Ohno was no exception. It took Ohno about twenty-five years to push his production system through all of Toyota and its suppliers' facilities. The deployment process at times was slow. Even when it was deployed at the Georgetown, Kentucky, facility, it took about ten years to produce to capacity, whereas Ford's plant in Highland Park, Michigan, scaled to capacity almost instantly.[6]

In the United States, industry had the advantage of an extensive population with a large demand and the ability to create a strong business model based on economies of scale. Toyota had a much more diverse market with a much lower volume in demand.[7] Ohno, despite this disadvantage, created a production system that was more efficient than one based on economies of scale like in America.

Just-in-time was the central idea that Ohno tried to optimize as part of TPS. One of his key metrics was lead time.[8] The shorter the lead time, the more optimized the just-in-time production system. This optimization was achieved through the continuous flow of products, which moved forward while information flowed backward, signaling the replenishment of products pulled downstream. Lead time was drastically reduced because products were consistently moving, not sitting in inventory. Timelines shrank through the elimination of nonvalue-added tasks.[9]

Ohno needed his team to think in different ways. He was quoted as saying, "common sense is always wrong."[10] This kind of thinking helped Ohno challenge common sense and the accepted practices to develop new, more innovative ways for producing automobiles.

The challenge of common sense is that sometimes we do not question the things we should. I was asked to perform an evaluation of a new replacement machine. Our current machine had four spindles that could hold four parts. The four spindles turned ninety degrees each cycle. The first spindle position was the load and unload station. The second position was a wait station. A nozzle sprayed a chemical compound at the third position. The fourth position was a cooldown station.

The vendor offered an upgraded machine with six spindles and claimed it was more efficient than the machine with four spindles. The process within the new machine was the same except that there were now two wait slots and two cooldown slots.

This second cooldown slot presented a problem to our process. For our normal process, tape was placed on the product to prevent spraying on unwanted areas. The tape left a residue on the part that needed to be cleaned. If the part was warm, cleaning was simple. The second cooldown station in the new machine waited too long, making it difficult to clean off the residue. In addition to the lengthier clean times, the overall process time in the machine increased by 50 percent because now there were 50 percent more spindles. Upon further investigation, it was deemed that six spindles were less efficient than four spindles.

Another "common sense" issue I run into constantly within health care is the idea that adding machines in the sterilization process department (SPD) increases capacity and resolves constraint issues. That is not always the case because the manual work is sometimes not evaluated. If the staff is run ragged trying to perform their duties to include loading and unloading machines, then adding new machines that don't address the manual work does not necessarily improve the capacity. Yes, the machine capacity increases, but the people capacity can potentially decrease. Sometimes, common sense requires more investigation.

Like Shingo, Ohno felt statistical sampling was wrong-minded and believed that 100 percent inspection could be achieved by employing *jidoka* and *poka-yoke* techniques in all production steps. This belief created some problems for Ohno with the US occupation headquarters while they were present in Japan because they did not necessarily agree with Ohno. They favored Deming's approaches.[11]

The evolution of the TPS from the 1940s through the 1970s was very closely linked to Ohno's career. His success was heavily due to the leadership support of Toyota's president, Eiji Toyoda.[12]

Let's take a look at Ohno's history and the evolution of the TPS through the decades.

1930s

The 1930s was a time in which Ohno was likely introduced to the just-in-time production system that was being developed by Kiichiro Toyoda. Ohno noted that Japanese production was far less efficient than other countries. He began creating systems that ensured production work was repeatable.

Taiichi Ohno graduated from the Nagoya Technical High School and subsequently joined the Toyoda Spinning and Weaving Company in the spring of 1932. He became very familiar with Sakichi Toyoda's legacy, though Sakichi died two years earlier.[13] Just-in-time production began in the 1930s when Kiichiro Toyoda created a new production method that addressed lower production volumes with greater diversity and accommodated a lack of resources, such as space.[14]

In 1937, while working at the Toyoda Spinning and Weaving Company, Ohno learned that German workers were three times more efficient than Japanese workers, and that American workers were three times more efficient than German workers. After a quick calculation, Ohno was surprised to learn that American workers were nine times more productive than Japanese workers.[15]

Sometime between 1937 and 1938, Ohno was tasked with producing standard work for textile production. He accomplished this task by studying a book bought from Maruzen.[16] This would be the start of the standard work principle as a visual control mechanism to manage the TPS.[17] "Standard work combination" was a way to address complexity by combining material, workers, and machines in the most effective and efficient manner. The procedure listed three critical elements:

1. Cycle time, the time required to complete a cycle of work.
2. Work sequence, the sequential order of the work.
3. Standard inventory, the minimum amount of inventory needed at each step to ensure work did not stop.[18]

1940s

Ohno worked in textiles until 1942 when the Spinning and Weaving Company was dissolved. In 1943, he was transferred to the Toyota Motor Company, where he embraced the idea of *jidoka*. He believed *jidoka* could be applied to automobile production in the form of autonomation, or automation with a human touch.[19] After World War II, Toyota's future was bleak because it had no money and was indebted to banks. Productivity was only one-ninth of the Detroit automotive companies.[20] When Ohno came to the automotive side of the business, it ran similarly to the way Ford ran its factories, but Ford was able to take advantage of economies of scale whereas Toyota could not; the demand for large batch sizes were not constant.[21]

August 15, 1945, marked Toyota's new beginning, and it was about this time when Toyota's president, Kiichiro Toyoda, stated his goal: "Catch up with America in three years; otherwise, the automobile industry of Japan will not survive."[22] One of the objectives was to produce high-quality goods.[23] After World War II, products from America started to come to Japan, including chewing gum, soda, and cars.[24] This spurred the need for Japanese competition by producing high quality goods.

Ohno was in charge of the Koromo plant in 1947. To meet the president's edict and to improve the productivity of workers, Ohno organized the machines so one operator could handle numerous machines. The machines were arranged in the shape of an L or parallel to each other. The goal was simply to make it so that one worker could be productive across a range of machines.[25]

Ohno was promoted to manager in 1948 and was put in charge of Toyota's engine manufacturing department.[26] It was here he noted the shop was consistently waiting on parts due to batched production. He also noted that workers spent too much time watching their machines and that defective parts were only discovered when they reached quality control stations. He started to adapt Sakichi Toyoda's *jidoka* concept in the machine shops by installing go/no-go gauges and limit switches. In this way, the machines stopped working as soon as an error was detected.[27]

Ohno discovered that when too much inventory was present, there was always one part missing. He worked with his production lines to not produce more than what the previous step withdrew, a precursor to *kanban*.[28]

Ohno also reorganized machines into cells versus process villages. This helped minimize the transportation of parts throughout the production process. These cells, generally U-shaped, grouped workers into teams and empowered them to collaborate and determine the best way to perform their operations.[29] Absent worker positions could be temporarily filled by team leaders.[30]

In 1949, inflation was rampant in Japan, so the American occupation introduced the Dodge line to counter it. About this time of macroeconomic problems, the American occupation also restricted credit, possibly a bit too much. This contributed to the depression in Japan, and Toyota started to suffer financially.[31]

1950s

The 1949 crisis continued into the following year when Kiichiro resigned after accepting responsibility for the layoff of a third of Toyota's workforce. Afterward, jobs were guaranteed for life. Ohno's methods were to become the norm, but layoffs would not be allowed because of process improvements from that moment forward.[32]

In June 1950, the Korean War broke out, and Toyota received a large order to make trucks for the US Army. Having crawled out of their financial problems, Toyota learned how to increase production of automobiles without significantly increasing head count.[33]

Eiji Toyoda, the cousin of Kiichiro Toyoda, traveled to the United States for a three-month tour, where he visited various plants. He was surprised to learn that production systems remained relatively the same as in the 1930s.[34] Eiji studied the Ford River Rouge plant intensively, and after discussing his observations with Ohno, determined that mass production would not work for Toyota. The Rouge plant was considered the largest, most efficient manufacturing location in the world at that time.[35]

Eiji Toyoda assigned Ohno a daunting initiative: catch up to Ford's levels of productivity. Ohno studied Henry Ford's book, *Today and Tomorrow*. He learned the importance of continuous flow from Ford's book but needed to innovate a way to achieve continuous flow without producing large batches of material, which would tie up cash.[36] This began Ohno's journey to apply and integrate both *jidoka* and one-piece flow into the just-in-time production system. The ideal batch size would always be one.[37]

In the early 1950s, Ohno visited a supplier and observed a worker do nothing but watch a machine perform lengthy tasks. The machine never broke down, yet the worker continued to watch the machine. Ohno thought to himself, "What a terrible waste of humanity."[38] Ohno wanted to eliminate this terrible waste at Toyota, so he arranged various machines to be in sequence with each other so that one would produce a part and move it to the next station without inventory in between. The person worked in tandem with the machines. This represented a radical change for the conventional mass production system of the day.[39] This setup also maximized both the strengths of workers and machines and was certainly not a waste of humanity.

Ohno aggressively worked on pull systems to optimize the just-in-time system at Toyota, and he had several inspirations. In the 1950s, the first US-type supermarkets appeared in Japan.[40] By 1953, from Ohno's study of these supermarkets, the first manufacturing supermarket system was adopted into Ohno's machine shop.

With American supermarkets showing up in Japan, the influence of Kiichiro Toyoda's just-in-time concept, and literature such as a 1954 article about Lockheed saving $250,000/year through the use of a supermarket system, Ohno had ample references to build Toyota's production system.[41] He strived for a one-piece flow system, but this was not realistic in every case. Therefore, he created small stores of parts that were replenished upstream when used by a downstream process, but only in the quantities used.[42]

By 1955, Toyota's focus shifted from only producing high-quality goods to also producing those high-quality goods in the exact quantities needed.[43] It was at this time Japan entered into a high-growth period.[44]

In 1956, Ohno traveled to the United States, where he toured General Motors and other manufacturing facilities. It would be the US supermarkets that impressed Ohno the most. According to sixsigmadaily.com, one of the supermarkets Ohno visited was Clarence Saunders's Piggly Wiggly. Ohno admired the way merchandise was replenished on the shelves in an efficient and timely manner.[45]

Ohno needed a way to communicate upstream items that were needed downstream, and with that came the *kanban* system. *Kanban* was basically translated as being a signal.[46] In 1953, Ohno introduced *kanban* cards to help improve the flow of information and facilitate the flow of products throughout the process. The information flowed backward, giving previous steps permission to produce parts once pulled, while production moved forward at an equalized rate.[47] Process sequencing, use of standard work, and the pull scheduling system had already been introduced to TPS in the 1950s.[48] These concepts began to meld together to create a superior production system.

I previously spoke about the production supermarket system in one of the manufacturing plants I worked in. Over time, we found that our production supermarket was not perfected. Though we were following all our rules of the supermarket system, we missed production numbers regularly across all shifts because parts did not replenish the supermarket quickly enough.

Working as a night supervisor, I felt it was my responsibility to hit the production targets, and if there was a problem, I needed to rectify the situation. The concept to me was simple. If I had the raw material, the ability to prepare it for final assembly, and the final assembly schedule, then I should be empowered to complete the production schedule.

Each night I brought my two shift leaders together. We scrubbed the production schedule to determine if the material was available and if not, determine if we could make it available. This was good at first, but we needed something more.

Prior to the start of our shift, I tasked a team member to conduct an inventory of products at each production step. In parallel, my assembly lead reviewed his schedule and gave me a list of what products he needed. From there I created an algorithm, using computer spreadsheets, to determine what to pull through the system. This spreadsheet had inputs from both the assembly schedule and the preshift inventory. Once the spreadsheet calculated the production needs, I met with my manufacturing shift leader, who made sure those products were prioritized. Our production consistently ranged between 99 percent and 104 percent completion rate.

Eventually, leadership noticed our performance, and I was invited to a meeting to discuss this issue. Several people were present to include scheduling, continuous improvement, management, supply chain, and other shift leads.

At the meeting, it was stated we needed to know what to prioritize throughout our manufacturing process. I offered up my spreadsheet, which generated excitement. Immediately the supply chain personnel, schedulers, and the continuous improvement team members discussed ways to automate my process.

Within a few short weeks, they automated a *kanban* system that augmented the supermarket system, making it more flexible to ever-changing demands. All shifts embraced the new *kanban* system immediately, and production skyrocketed. The *kanban* system was also spread to other locations.

The leader of the continuous improvement team called me to his cubicle to show me the final report. I was surprised at the use of statistics, the order of the report, and the simplicity of the message. I was more surprised by the cover sheet. I was given full credit for the project. I told him I was not the one who did all this work. He said something to the effect of, "Everyone knows that, but it was you who cracked the code, and it was your work that served as the blueprint for everything we did. You deserve the credit."

This was my very first Lean project—using *kanbans*—and it worked! How lucky I was to be associated with such amazing people. One more note: true Lean systems don't have to prioritize much.

1960s

As stated earlier, Nissan stunned Toyota when it won the Deming Prize in 1960, which served as a motivator for Toyota to ramp up its production system by infusing total quality control with Ohno's ideas.[49] Toyota struggled in the 1960s because its automobile division had a significant inventory of cars: three years' worth. Though there were major quality improvements, these improvements alone could not solve Toyota's financial problems. The company tried to achieve just-in-time production, but it wasn't yet successful. Ohno continued pushing *kanban* as a way to signal and authorize manufacturing of product replacements upstream when pulled downstream. This process helped Toyota make great strides toward true just-in-time production, and it was able to produce vehicles that would sell quickly and not tie up cash flow in inventory.[50] In 1962, the *kanban* system was deployed company-wide when Ohno was able to put it in the forging and casting departments, completing a ten-year *kanban*-implementation journey.[51] Toyota won the Deming Prize in 1965.

Just-in-time was not yet fully achieved by the late 1960s. Another major hurdle remained for Ohno. The changeover time of his larger machines was still too long, which resulted in large batches of in-process inventory. Ohno learned that the press changeover times in Germany's Volkswagen plants were half that of Toyota's. Ohno attempted quick changeovers as early as the late 1940s, with little success.[52] With the help of Shigeo Shingo and the immense efforts by those associated with Taiichi Ohno, such as Kosuke Ikebuchi, Toyota was able to get the changeover of a 1,000-ton press down to three minutes. As noted earlier, Shigeo Shingo wrote of his gratitude to Ohno for setting such a challenging goal, stating that this was the birth of SMED.[53] By producing smaller batches with the quick changeover, Ohno saved money in his production system.[54] Carrying costs of large inventories could be eliminated, and stamping mistakes could be caught and addressed immediately.[55]

By the mid-1960s the main Toyota facilities were using Ohno's enhanced just-in-time methods. The next phase of spreading TPS involved working with Toyota's suppliers. Suppliers still had large finished goods inventories, and they could not respond quickly to Toyota's continual changes in production requirements. Toyota identified forty-two suppliers.[56] In 1969, Ohno established the Production Research Office to work with Toyota's suppliers.[57]

1970s

By 1971, Ohno proclaimed the TPS had been accomplished.[58] A couple of years later, the value of that proclamation was put to the test.

The oil crisis in 1973 served as a wake-up call to many industries. Toyota appeared to be faring well compared to other automotive companies. The need for a leaner production system was apparent, and like a good citizen, Toyota started teaching the *kanban* system to other companies.[59]

Toyota was not immune to the effects of the oil crisis. Demand decreased, so the company had to reduce production by 1974.[60] Due to improvements inherent within TPS, the operating lines could be operated by one or more operators, giving it the flexibility to reduce production with relative ease.[61] Toyota's small-car production and the leaner production systems helped Toyota navigate the troubled waters of that era.

It would not be until the late 1970s that many of the TPS concepts were written down.[62] Ohno published his book, *Toyota Production System: Beyond Large-Scale Production*, in 1978. It became available in English about ten years later.[63]

On my third trip to Georgetown, Kentucky, I asked an instructor which book I could study to learn about TPS. I shared with him a list of books I was studying. He scoffed and dismissed each one on my list. He then pulled out his heavily marked and dog-eared copy of Ohno's *Toyota Production System*. I immediately purchased a copy. I felt like I was drinking through a firehose during the first 100 pages.

Before his book, Ohno wrote down very little, if any, of the details of TPS. Hence, much of the work was hands-on, including training. Ohno realized that this may have been a reason why TPS did not spread across the Toyota facilities and suppliers very quickly.[64]

He spent much of his time in the 1970s working with the first- and second-tier suppliers to convert them to TPS, which was largely completed by 1978. The need to record the elements of TPS may have been apparent to him at that time.[65]

1980s

In a 1980 interview, Ohno stated that Ernest Kanzler, one of Ford's managers, developed the concepts of reducing inventory through careful planning and coordination of raw materials when introduced into the factory. This was one of the influences Ohno had used to create a just-in-time system.[66] This was another example of Ohno's persistence. He pulled in anything he could from any source he felt appropriate and integrated it into TPS, which operated as a well-oiled machine.

Ohno's work could be compared to the work of Frederick Taylor. Taylor focused on separating planning from production, whereas Ohno put them back together again. Like Taylor, Ohno wanted to achieve a humane and harmonious workplace. He used many of the scientific management techniques such as time and motion studies and standardized work.[67]

With the help of Shigeo Shingo, Ohno demonstrated that high cost was not necessary to create high levels of quality.[68] Ohno, by his own admission, stated these tools could be applied beyond manufacturing.[69] Years later, that statement was realized when the concepts of TPS were called Lean, and principles were developed to help apply these concepts to other industries.

OHNO'S ANTICS

Taiichi Ohno behaved in a manner some may have considered unorthodox. Some people referred to him as a genius. Though he may very well have been a genius, there were three descriptors that could also describe him. He was hands-on, persistent, and obnoxious.

Though Ohno was obnoxious, he had Eiji Toyoda's support.[70] This support was likely the reason his obnoxious behavior was tolerated. Pupils of Ohno often could not remember receiving compliments from him, but they remembered the regular and constant tongue lashings and criticisms.[71] Below are some accounts that have made the rounds.

- Ohno had his engineers draw a circle on the plant floor and then told them to stand there for up to eight hours! At the end of the day, Ohno simply told them to go home. Upon reflection, the engineers observed many process problems and learned the power and importance of deep observation and thinking for oneself.[72]

- Ohno fired and hired people several times in a day. He would rant, throw chairs, and send engineers to a different, difficult location with the promise not to come back if improvements were not made. Their stories reached a higher level of honor as the scene of the incident was made more public.[73]

- Ohno believed the activities on the shop floor served as a reflection on management.[74] In the 1980s, Taiichi visited a foundry at the Showa Manufacturing Company. After a quick observation, Ohno had the company president bring the plant manager to him. Ohno told the plant manager that he was incompetent. He told the president to fire the plant manager immediately. This was the start of a productive relationship involving all three men. The plant manager was not fired; instead, he embraced and implemented Ohno's ideas.[75]

- Ohno tended to try ideas on machines while alone at the plant, apparently thinking that nobody was aware. He would test out his concepts, but in the process, he created a significant number of defects that he threw into a pond out back. As the author Jeffrey Liker put it: "After a time, the pond filled up with his rejects (or so goes the legend)."[76]

- Ohno admitted to extreme behaviors, such as yelling at others. "I could yell at a foreman under my jurisdiction, but not at a foreman from the neighboring department," he said.[77]

- Ohno was able to produce 5,000 units (Corolla engines) in 1966 with only eighty workers. He asked the head of the engine section how many workers were needed to make 10,000 units (double the amount). The head simply doubled the people and said, "one hundred sixty." Taiichi yelled, "In grade school I was taught that two times eight equals sixteen. After all these years, do you think I should learn that from you? Do you think I'm a fool?" In the end, he only needed 100 workers, not 160, to produce the 10,000 units.[78]

One may have assumed from these stories that Ohno was a difficult person who was very hard on his direct reports. He was, but Ohno valued his workers and noted they were essential resources. Ohno and his team regularly involved the workers in improvement initiatives.[79]

TOYOTA PRODUCTION SYSTEM

With the help of many, Ohno was able to successfully align and integrate *jidoka* and just-in-time processes into one production system. He improved upon the just-in-time system with continuous flow that strived to achieve one-piece flow. He created cells in which large machines were miniaturized and positioned next to each other so products did not sit in inventory. With the help of Shigeo Shingo, he reduced the downtime of large machines by quickening the changeovers so machines could change over more often and prevent the buildup of inventory.

Manufacturing did not turn on a dime, so schedule leveling, or *heijunka*, was deployed to create consistency in the manufacturing process. Another tool used was an *andon* line stopping mechanism that empowered workers to stop the line and prevent moving defects downstream. When the *andon* chord was pulled, the worker pulling the *andon* was immediately assisted by others to remediate the respective problem. In most cases, the assembly line did not stop. The *andon* served mostly as a signal to ask for assistance.

There were many other contributors to TPS. Hiroyuki Hirano was credited with the 5S system, Seiichi Nakajima was credited for Total Productive Maintenance (TPM), and Kenichi Sekine was credited for continuous flow. Let's not forget Shingo's influence with expanding *jidoka* and source inspection and SMED.[80] Ohno also employed the use of Five-Whys, a root cause analysis tool, in his problem-solving analysis.

This new production system had many benefits. Rework was reduced continually, and the quality of the automobiles improved significantly. Why? Because Toyota was not relying on end-of-line inspections that could not realistically catch every possible issue.[81] Quality was built into the production system. When all the tools were in place connecting all departments together, the production system truly acted like the nervous system of the human body. Similarly, TPS required constant care.

Indeed, just-in-time, which was often synonymous with TPS, was described by Norman Bodek in this way: "The automatic nervous system responds even when we are asleep. The human body functions in good health when it is properly cared for, fed and watered correctly, exercised frequently, and treated with respect."[82]

Here are some of Ohno's highlights while creating TPS:

SUPERMARKETS IN MANUFACTURING

Ohno deployed the use of manufacturing supermarkets in TPS. A downstream process pulled product from a manufacturing supermarket, which in turn signaled the need for replenishment by the upstream process. This became known as a pull system.[83]

KANBAN

This signal system was the information tool sent upstream to authorize the replenishment of products downstream.

AUTONOMATION

Ohno spent many hours after work on machines to prove his ideas on autonomation, or the building of human intelligence into machines. He encountered resistance on the shop floor, but he persisted. In the end, he produced a system that permitted workers to work on several machines at the same time, performing load and unload tasks.

WASTE

Ohno identified seven forms of waste, also known as *muda*, as activities that consumed resources but did not create value. These activities included mistakes requiring rectification (defects), production of items not wanted (overproduction) resulting in a pileup (inventory), processing steps not needed (extra processing), movement of employees (motion) and movement of goods (transport) without purpose, and people standing around waiting for an upstream process to deliver items (waiting).[84] Overproduction was referred to by Ohno as a crime.

Mass production required additional investments in building maintenance, warehouses, additional workers, machines, parts, materials, energy, oil, electricity, forklifts, tow trucks, pallets, skids, and interest payments on loans. Problems were also hidden in the inventory.[85] Ohno also identified *mura*, often translated as inconsistency/variation/unevenness, and *muri*, or unreasonableness/overburdening.[86] Value-added work was some form of processing, or the changing the shape/character of a product. The better the ratio of value-added tasks to non-value-added tasks, the better the work efficiency.[87]

PART III

1973-TOMORROW: THE OIL CRISES TO TOMORROW AND BEYOND

After World War II, the US automobile industry started to slowly scald itself in a pot of hot water and did not realize the water was boiling until the oil crises of the 1970s. There were several factors that contributed to this situation. Following the war, the United States was the only economic power in the world, but US manufacturing and production quality started to decline. Additionally, the market was different in the United States versus other parts of the globe.

The demand for larger vehicles increased in the United States, and American automobile companies positioned themselves to satisfy this demand. Inventory had a high likelihood of selling. There was no need for smaller cars because gas was cheap. All that would change with the oil embargos of the 1970s. It was during the first oil crisis that the world recognized there was something different about the Japanese approach to manufacturing compared to the United States.

The oil crises throughout the 1970s increased demand for smaller cars in the United States quickly. The need for improved, less wasteful manufacturing was elevated. Foreign car companies such as Toyota not only increased their sales to American consumers but also began manufacturing in the United States using American workers.

Could American workers function in a Lean system, as represented by TPS? The answer was a definite *yes* as demonstrated with the GM/Toyota partnership in California, called New United Motor Manufacturing, Inc. (NUMMI), and subsequently followed by the opening of Toyota plants on US grounds.

CHAPTER 11

A TIME FOR CHANGE
(1970s)

*"Too much past success, a lack of visible crises,
low performance standards, insufficient feedback
from external constituencies, and more all add up to:
'Yes, we have our problems, but they aren't that terrible
and I'm doing my job just fine,' or 'Sure we have big problems,
and they are all over there.' Without a sense of urgency, people
won't give that extra effort that is often essential."*[1]
–John Kotter

OIL EMBARGO

CAR OWNERSHIP

Several things contributed to Toyota's success after 1950. First, with the outbreak of the Korean War, Toyota won a contract to produce trucks for the United States. Japan also became a car-owning society in which passenger cars proliferated from the 1950s to the 1970s. The Japanese government imposed a tax system that favored smaller cars, and thus lightweight-car production increased in Japan.[2] These smaller cars were cheaper to make, which created a demand for a greater variety in the types of cars, a condition the TPS was designed to accommodate.

After World War II, most of the American buyers wanted larger, high-powered vehicles that used a significant amount of fuel, a condition that was amicable for US auto manufacturers. Conditions in the United States made it easy for car owners to have such vehicles because of low gas prices, an excellent highway system, and the significant traveling distances.[3] It was not profitable for US automakers to produce small, fuel-efficient vehicles, so they were willing to allow foreign producers to supply this market, most notably Europe and Japan.

In 1957, small and inexpensive European cars achieved some noteworthy sales in the United States when America entered a recession.[4] By 1959, the Volkswagen Beetle led the way of imports to the US market. These cars, though, were underpowered, frequently overheated, and often just broke down.[5] In response, Detroit automakers reduced the number of imports with the introduction of smaller vehicles such as the Ford Falcon, Chevrolet Corvair, and Plymouth Valiant. Within ten years, in the late 1960s, these vehicles grew both in size and weight, and by 1968, imports of small vehicles increased again.[6]

North American imports increased significantly between 1955 (approximately 92,239 imports) and 1970 (approximately 1,753,739 imports) while Japan's exports to the rest of the world increased from six in 1955 to 235,737 in 1970.[7] This trend continued into the 1980s.

OIL CRISES

Between 1950 and 1973, the Middle East's world oil production rose from 7 percent to 42 percent. Oil prices were already rising by 1973. About ten days after the outbreak of the Arab-Israeli War in October of 1973, the Organization of Arab Petroleum Exporting Countries (OAPEC) cut oil production to force Europe and Japan to pressure the United States to change its policy regarding the Arab-Israeli situation.[8] By March, the United States helped to negotiate cease-fire agreements, so OAPEC ended the embargo.[9]

This oil crisis impacted Japan's economy and thus ended a high growth era. Several analysts believed that Japan would not be able to grow its auto industry because of the increased gas prices and decreased car demand.[10] The opposite occurred. In the case of Toyota, their just-in-time strategy minimized the amount of dollars tied up in hard-to-sell inventory. Also, Toyota produced smaller, more fuel-efficient cars and was able to respond quickly to the demand shift.[11]

Taiichi Ohno, in his book *Toyota Production System: Beyond Large-Scale Production*, stated it was not until the 1973 oil crisis that attention was given to the TPS by the broader Japanese industry and that Japanese managers noticed Toyota's success with their pursuit to eliminate waste.[12] John Shook equated this to the TPS efforts to direct all activities to make money.[13] This was a statement that closely related to Eliyahu M. Goldratt's book, *The Goal: A Process of Ongoing Improvement*, in which the goal was to make money. The book introduced the world to the theory of constraints.

Surgeons and other stakeholders commonly ask me which book will help them learn quickly about patient flow. The book that tends to resonate most is Goldratt's book, *The Goal*. Many surgeons tend to understand constraints.

When evaluating OR efficiency, I calculate the capacity of the SPD. The instrument tray capacity is normalized to the number of patients the SPD can support.

Let's assume an OR has a demand for 5,000 patients, and the SPD has a capacity for 15,000 trays. That is not enough information to determine if the SPD has the capacity to meet the demand. If 15,000 trays correlate to only 3,000 patients, then it is easy to see that the SPD capacity cannot support 5,000 patients.

Goldratt's book is far more involved, but the idea of the constraint is easily understood. In my experience, several people, especially surgeons, can use *The Goal* to think in new ways and become open to differing ideas.

In one hospital, I was able to show that the SPD only had 50 percent capacity to demand. I helped the surgeons understand they could directly reduce the burden on SPD by simplifying and standardizing their trays.

Some surgeons who read *The Goal* took proactive measures to help the SPD. The surgeons standardized their trays, reduced the number of instruments in each tray, and reduced the total number of trays needed for surgery. Soon, some SPDs had the ability to meet the demand.

By 1975, more than half the leading industrial companies were in a recession,[14] and by the end of 1979, 60 percent of cars sold in the United States were either compacts or subcompacts.[15] These smaller cars offered better fuel efficiency and were of good quality at competitive prices.[16]

The Iranian Revolution disrupted oil supplies again in November 1978 and early 1979, only to resume at reduced levels.

Production was increased elsewhere in the world, but oil prices still more than doubled.[17]

Gas prices increased considerably between January 1979 and March 1980 (80 percent), so the American consumer converted to the more fuel-efficient, smaller, imported cars.[18] Foreign car imports threatened the survivability of the American auto industry.

The US auto industry embarked on several tactics to survive, such as Chrysler receiving $1.5 billion in loans from the federal government in 1980. Ford and the United Auto Workers (UAW) attempted to seek relief from the increasing number of imports by filing a trade safeguard case in 1980 but were denied by the US International Trade Commission. GM increased automation in plants to decrease costs, but the opposite occurred—their costs increased.[19]

American demand for smaller, lighter, and more fuel-efficient cars began to rise, but US automakers could not provide these cars effectively. This opened the window for Japanese automakers to enter the market with cars that were more diversified, provided more comforts, and produced less pollution.[20]

US automakers were forced to close some plants. Between 1945 and 1960, twenty-five plants were opened, and none were closed. Four opened and four were closed between 1960 and 1978. Between 1979 and 1991, eight plants were opened, but twenty were closed.[21] General Motors tried to close some of their northern unionized plants in the 1970s in an effort to open plants farther south with a goal of reducing labor costs. The UAW successfully unionized those plants, so the GM strategy was not successful.[22]

FOREIGN PRODUCTION ON US GROUNDS

Japan's automotive industry had several things going for it. A production system that produced less waste helped to improve its financial strength by 1973, but other factors also came into play. Research and development (R&D) had a close relationship with the manufacturing process[23] and a strong working relationship with suppliers, especially in developing new products.[24] US automakers of the 1950s and 1960s, on the other hand, were still dependent on 1930s technologies, with the last major innovation being the 1940 automatic transmission in the GM Oldsmobile.[25]

Japanese automakers took advantage of low wages, lower costs for imported materials, and favorable trade positions.[26] The demand in their local markets was for smaller, more fuel-efficient vehicles so when the oil crisis occurred, companies such as Toyota had the supplies and systems to enter the US market.

The Big Three—General Motors, Ford, and Chrysler—suddenly had to decide whether to compete, invest, or partner with European and Japanese automakers, especially as Japanese companies laid plans to establish a foothold in the United States with manufacturing facilities of their own. Honda was the first in 1982, followed by Nissan in 1983, to open plants in the United States.[27]

By 1981, Japanese vehicles accounted for 41 percent of the world's exports of assembled motor vehicles. It was the only nation that manufactured more vehicles in 1981 than in 1973.[28] It was about this time that Detroit motor companies stated preferences to Japanese investments versus strict competition from Japanese imports.[29] One such example of a joint investment that came to fruition was the GM/Toyota joint venture known as NUMMI in Fremont, California.

Foreign trade zones (FTZ) in the United States allowed for the importation of parts that were duty-free. Payment of the duty was deferred only when exporting the final product. Free trade subzones (FTSZ) were permitted in 1952. These FTSZs were manufacturing plants and were not necessarily within an FTZ. It was common that an automotive plant was an FTSZ.[30] This created a favorable position for foreign automakers to invest in efforts to manufacture in the United States.

Other Japanese car companies, such as Honda, Nissan, Toyo Kogyo (Mazda), and Mitsubishi also found ways to grow the US market through independent investments and joint ventures.[31] After NUMMI, Toyota built its own successful independent plant in Georgetown, Kentucky. The United States became a manufacturing home for foreign automobile companies.

CHAPTER 12

LEAN PRODUCTION BY AMERICAN WORKERS
(1984–2010)

"We never make for the sake of making."[1]
–HENRY FORD

NUMMI

General Motors sought a leaner, more efficient production process. Toyota was a viable partnership. Both organizations could benefit; thus, New United Motor Manufacturing, Inc. (NUMMI) was formed.

THE BEGINNING

General Motors owned a plant in Fremont, California. Between 1962 and 1982 the plant made light trucks and cars, but it did not perform well in the areas of quality, safety, and productivity. Absenteeism was common, so the plant was shut down in 1982.[2] Shortly after, Roger Smith, GM's CEO, approached Toyota about a joint venture using the Fremont plant.

Toyota and GM negotiated an equal equity joint venture in the assembly of small cars at the GM plant located in Fremont. Toyota contributed $100 million; GM contributed $11 million cash and the $89 million Fremont plant. Together, they raised $350 million for a stamping plant.[3] The first automobile at the NUMMI facility was produced in December 1984.

General Motors benefited by learning first-hand about TPS, and Toyota learned how to apply TPS in the United States using an American workforce.[4] Together, the plant produced small cars, helping GM's fuel-economy ratings.[5]

Toyota used the GM opportunity to gain a foothold in the United States and in two short years opened the Toyota plant in Georgetown, Kentucky. Expansion in the United States was not Toyota's only motivation to partner with GM. The company wanted to help the United States by creating high-paying jobs for Americans.[6]

Toyota senior managers led the plant and implemented TPS.[7] One of the successes of NUMMI involved a new approach to working with unionized facilities, which were traditionally filled with mistrust of management.[8] Initially, there was a perception that TPS would work people harder by speeding up the line, and consequently, there were some hostile attitudes. Toyota, though, sent people to Japan for three weeks to see TPS at work, and this converted many who were originally skeptical.[9]

Trust by employees continued to grow for several years. In fact, during the 1987–1988 time frame, orders for vehicles were reduced, causing the plant to run at 75 percent capacity. General Motors remained committed to its workers by refocusing the excess employees to other activities as opposed to laying anyone off.[10]

DURING

By 1985, a GM study determined that NUMMI produced automobiles of better quality with about 40 percent fewer human resources and from 20 to 50 percent lower investments than other GM facilities of similar comparison.[11]

By the 1990s, without much supporting research, positive comments about the production system began to surface in the areas of how a humane approach was incorporated in the production environment. One person commented, "Work routines still resembled those of scientific management, but they allowed for a degree of democratic input."[12] Another person stated optimism "for a more human and participatory work environment."[13]

By 2002, according to Harbour Report's annual auto-plant productivity report for North America, GM beat Ford. Also, in 2002, GM achieved recognition from JD Power and Associates for low customer complaints in the first ninety days of car ownership. In 2003, NUMMI produced 395,000 vehicles and by 2004 employed about 5,700 workers.[14]

General Motors began improving quality and productivity performance throughout the organization by transferring knowledge from NUMMI to other GM facilities.[15] Lean manufacturing techniques started to be adopted throughout GM. Gary Cowger, GM vice president of manufacturing and labor relations, stated that GM learned from Toyota and that GM had "to give credit where credit is due."[16]

Practically every minute at the NUMMI plant was structured, making it very mechanistic. With that in mind, it appeared that NUMMI attained what scientific management tried to do.[17]

THE END OF A PARTNERSHIP

The GM/Toyota relationship deteriorated, and eventually GM had financial difficulties. There were several causes for GM's financial problems. One of the major causes was the excess legacy costs related to retirements and health care. There were also problems in sales and marketing.[18] The Fremont plant began to operate at only about half capacity and started to lose money. The end was near. On April 1, 2010, the NUMMI plant was shut down, and the GM/Toyota joint venture officially ended when the last Toyota Corolla rolled off the line.[19]

CONCLUSION

"So far as I know I never gave in to the temptation to tailor facts to ideas rather than adapt ideas to facts."[1]
—ALEXIS DE TOCQUEVILLE

WHERE DID LEAN COME FROM?

John Krafcik, an American engineer hired at NUMMI, spent time in the Japanese Toyota factories learning the TPS. As part of his MBA research, John surveyed ninety auto assembly plants throughout the world, and it was Krafcik who used the term "Lean" to describe TPS. He said Toyota's production system used about half the human effort, manufacturing space, tool investments, and new production engineering hours than mass production systems.[2]

When James P. Womack, Daniel T. Jones, and Daniel Roos authored *The Machine That Changed the World* in combination with the term "Lean," a new application of TPS emerged. Several books by such authors as Jeffrey K. Liker showed the application in different types of manufacturing settings. Womack and Jones, in their book *Lean Thinking: Banish Waste and Create Wealth in Your Corporation*, identified the five principles of Lean: (1) identify value, (2) create the value stream, (3) flow the value through the value stream, (4) pull to the value stream, and (5) pursue perfection. This thinking gave the world a foundation to apply variations of Lean concepts across multiple industries and settings.

CRITICISMS

Lean has drawn much criticism for various reasons.

1. Some argue that not every organization is an automotive company. True, but success has been achieved in numerous industries.

2. Some people say, "Only the Japanese culture could make it work." This is not true because many of the concepts originated in the United States, and there are numerous Americans today successfully working in Lean facilities.

3. Previous approaches focused on cutting head count, creating an acronym for Lean meaning "Less Employees Are Needed." It is true that Lean has a focus on increasing productivity with fewer resources, and it is true that market fluctuations, legislation, and other outside factors create a downturn in workforce demand. There were cases where companies did their best to make a commitment to not eliminate a person's job based on an improvement but failed to do so. Valuing people is critical to a successful Lean program. Ford and Ohno clearly valued workers. Even Taylor valued the worker. Training Within Industry's whole methodology was based on respect for the worker. Kiichiro Toyoda, as an example, resigned from his position when he could not maintain the jobs of his workers. Lean cultures value people.

There are more criticisms, but most arise from poor implementations, poor leadership, poor understanding, secondhand perceptions, and the like. It's poor reasoning to think there is something wrong with a piano because the piano player does not know how to play a piano. The same applies to Lean.

PURSUING A PERFECT LEAN SYSTEM

A perfect Lean system may have no waste whatsoever, but that can be difficult to visualize. What does a perfect Lean system look like? Is it a technology with the capability to replicate anything we want by asking for it like characters in popular science fiction shows?

But wait! Isn't it a form of waste to ask for what we want?

Wait again! Is the mechanism's very existence of occupying space when not in use also a form of waste?

What if we simply snap our fingers and whatever we think comes into being? Is the motion of snapping our fingers a form of waste?

This rendering may sound a bit ridiculous because it is. It is speculation as to what a perfect Lean system looks like. It's likely impossible to achieve. In the pursuit of the perfect Lean system, we may achieve greatness.

If you are thinking about beginning your Lean journey or have already begun, I offer this. We are always learning. There are resources, such as Toyota, that can help, but as situational leadership teaches, when we begin to learn anything, we simply do not know what we do not know. We need guidance.

Lean is not a fad. When all components of a Lean system are integrated effectively, it functions as a well-oiled machine. Maintaining this well-oiled machine takes considerable effort.

Lean is a progressive machine that will continue to evolve. It is not the solution to all problems. It is not the only way to produce, but the pursuance of perfection to create absolute value certainly opens the doors for Lean thinking.

Is there a principle of Lean that has yet to be discovered or recognized? Of course.

In writing this book, I tried to adapt my ideas to the facts so my research could be viewed as a genuine attempt to present ideas in a manner that fit the facts. I hope my respect for all the innovators and researchers surfaced clearly in this book.

Each of the innovators mentioned were competent and pragmatic thinkers. Some were calm like Dooley or more animated like Ohno. They may not have liked each other if in the same room. Despite their differences, there was one thing they had in common. They were persistent. Maybe that is the attribute of the person who will bring to the world something that will take our Progressive Machine to new heights.

Will that be you?

ENDNOTES

Chapter 1

1 Wren, 2005, pp. 81-82
2 Wren, 1998, Chapter 1
3 Schur, 2016
4 Wren, 1998, Chapter 1
5 Schur, 2016
6 History.com, 2018
7 Woodbury, 1960, p. 243
8 Wren, 1998, Chapter 1
9 Wukitsch
10 Mirsky, 1952, p. 301
11 Womack, 2011, p. 146

Chapter 2

1 Taylor, 1913, p. 6
2 Wren, 2005, p. 121
3 Wren, 2005, p. 125
4 Womack, 2011, pp. 281-282
5 Darmody, 2007, p. PS.15.3
6 Taylor, 1913, p. 7
7 Taylor, 1913, pp. 36-37: The original principles, word for word:
First. They develop a science for each element of a man's work, which replaces the old rule-of-thumb method.
Second. They scientifically select and then train, teach, and develop the workman, whereas in the past he chose his own work and trained himself as best he could.
Third. They heartily cooperate with the men so as to insure all the work being done in accordance with the principles of the science which has been developed.
Fourth. There is an almost equal division of the work and the responsibility between the management and the workmen. The management takes over all work for which they are better fitted than the workmen, while in the past almost all of the work and the greater part of the responsibility were thrown upon the men.

8 Wren, 2005, p. 128
9 Darmody, 2007, p. PS15.11
10 Walton, 1986, p. 9
11 Walton, 1986, p. 9
12 Meyers, 2002, p. 27
13 Meyers, 2002, p. 29
14 Shingo, 1987, p. xv-xvi
15 Dinero, 2005, p. xii
16 Dennis, 2007, p. 25

Chapter 3
1 Ford, 1926, p. 150
2 Mirsky, 1952, p. 272
3 Ford, 1926, p. 81
4 Ruttan, 2006, p. 21
5 Wren, 2005, p. 154
6 Wren, 2005, p. 263
7 Catalan Vidal, 2017, p. 5
8 Wren, 2005, p. 170
9 Parkes, 2015, p. 108
10 Parkes, 2015, p. 108
11 Wren, 2005, p. 264
12 Ford, 1926, p. 108
13 Ford, 1926, p. 109
14 Ford, 1926, p. 110
15 Ford, 1926, p. 110
16 Ford, 1926, p. 216
17 Ford, 1926, p. 231
18 Ford, 1926, p. 196. The Ford Principles are:
    1. Do the job in the most direct fashion without bothering with red tape or any of the ordinary divisions of authority.
    2. Pay every man well—not less than six dollars a day—and see that he is employed all the time through forty-eight hours a week and no longer.
    3. Put all machinery in the best possible condition, keep it that way, and insist upon absolute cleanliness everywhere in order that a man may learn to respect his tools, his surroundings, and himself.
19 Ford, 1926, p. 243
20 Eden, 1996, p. 511
21 Ohno, 1988, p. x

Chapter 4
1 Freeman, 2011, p. 61
2 Harkins, 2011, p. 146
3 Piggly Wiggly, 2011
4 Freeman, 2011, p. 14

5 Freeman, 2011, p. 14
6 Freeman, 2011, p. 34
7 Carden, 2011, p. 40
8 Carden, 2011, p. 41
9 Leonardo Group Americas, 2015
10 Liker, 2004, p. 18
11 Brucem, 2011, Russell, 2019, & Lopresti, 2017
12 Ohno, 1988, p. 26
13 Liker, 2004, p. 23
14 Carden, 2011, p. 40
15 Carden, 2011, p. 41
16 Schiffer, 2012
17 Ohno, 1988, p. 25

Chapter 5
1 Van Vliet, 2014
2 Magee, 2007
3 Liker, 2004, p. 16
4 Magee, 2007
5 Liker, 2004, p. 16
6 Ohno, 1988, p. 87
7 Wada, 2006, p. 96
8 Magee, 2007
9 Magee, 2007
10 Toyota Motor Corporation, 1998, p. 5
11 Toyota Industries Corporation
12 Concept drawing inspired by a pic from a blog by A. Sibaja
13 Wada, 2006, p. 91
14 Ohno, 1988, p. 88
15 Ohno, 1988, p. 89
16 Toma, 2017, p. 570
17 Wada, 2006, p. 104
18 Dennis, 2007, p. 96
19 Toyota Motor Corporation, 1998, p. 25
20 Toyota Motor Corporation, 1998, p. 26
21 Wada, 2006, p. 94

Part II
1 Robinson, 1993, p. 46
2 MacArthur, 1964, p. 352
3 MacArthur, 1964, p. 453

Chapter 6
1 Dinero, 2005, p. 29
2 Jacobs, 2002, p. 132

3 Jacobs, 2002, p. 136
4 Jacobs, 2002, p. 132
5 Jacobs, 2002, p. 132
6 Robinson, 1993, p. 37
7 Jacobs, 2002, p. 133
8 Dickens, 1945, p. 146
9 Dinero, 2005, p. 24; Robinson, 1993, pp. 37-38
10 Dinero, 2005, p. 92-93
11 Animated Knots
12 Jacobs, 2002, p. 135
13 Matsuo, 2014, pp. 227-228
14 Jacobs, 2002, p. 135
15 Jacobs, 2002, p. 134
16 Matsuo, 2014, p. 227
17 Robinson, 1993, p. 39
18 Dickens, 1945, p. 148
19 Robinson, 1993, p. 38
20 Soltero, 2011, p. 19
21 Gawande, 2009, p. 36
22 Robinson, 1993, p. 38
23 Soltero, 2011, p. 19
24 Dickens, 1945, p. 148
25 Robinson, 1993, p. 41
26 Keys, 1998, p. 129
27 Robinson, 1993, p. 42
28 Robinson, 1993, p. 41
29 Robinson, 1993, p. 38
30 Robinson, 1993, p. 43
31 Robinson, 1993, p. 37
32 Dinero, 2005, p. 26
33 Robinson, 1993, p. 37
34 Robinson, 1993, p. 38
35 Dinero, 2005, p. 23
36 Robinson, 1993, p. 39
37 Dickens, 1945, p. 147
38 Robinson, 1993, p. 40
39 Dickens, 1945, p. 149
40 Dinero, 2005, p. 11
41 Dickens, 1945, p. 147
42 Dickens, 1945, p. 146
43 Robinson, 1993, p. 38
44 Dickens, 1945, p. 150
45 Robinson, 1993, p. 40
46 Dickens, 1945, p. 145
47 Dickens, 1945, p. 145

48 Robinson, 1993, p. 44
49 Dinero, 2005, p. 41
50 Dinero, 2005, p. 41
51 Dinero, 2005, p. 42
52 Robinson, 1993, p. 35
53 Robinson, 1993, p. 35
54 Robinson, 1993, p. 36
55 Robinson, 1993, p. 46
56 Robinson, 1993, p. 36
57 Robinson, 1993, p. 46
58 Robinson, 1993, p. 47
59 Robinson, 1993, p. 47
60 Robinson, 1993, p. 46
61 Keys, 1998, p. 124
62 Keys, 1998, p. 127
63 Keys, 1998, p. 127
64 Keys, 1998, p. 129
65 Robinson, 1993, p. 42
66 Robinson, 1993, p. 48
67 Robinson, 1993, p. 51
68 Dinero, 2005, p. 48
69 Keys, 1998, p. 129
70 Dinero, 2005, p. xii
71 Keys, 1998, p. 131
72 Dinero, 2005, p. 48
73 Jacobs, 2002, p. 133
74 Jacobs, 2002, p. 131
75 Keys, 1998, p. 128
76 Keys, 1998, p. 129
77 Dinero, 2005, p. xii
78 Liker, 2004, p. 18
79 Robinson, 1993, p. 51
80 Dinero, 2005, p. xi

Chapter 7

1 Ohno, 1988, p. 91: Ohno believed this quote originated Kiichiro's just-in-time idea.
2 Liker, 2004, p. 17
3 Liker, 2004, p. 18
4 Liker, 2004, p. 79
5 Liker, 2004, p. 17
6 Toyota Motor Corporation, 1998, p. 5
7 Liker, 2004, p. 18
8 Toyota Motor Corporation, p. 5
9 Liker, 2004, p. 18

10 Toyota Motor Corporation, 1998, p. 5
11 Womack, 2011, p. 238
12 Luga, 2013, pp. 407-408
13 Liker, 2004, p. 79
14 Ohno, 1988, p. 80
15 Luga, 2013, p. 408
16 Ohno, 1988, p. 85
17 Liker, 2004, p. 18
18 Liker, 2004, p. 230
19 Womack, 2011, p. 315
20 Womack, 2003, p. 23
21 Liker, 2004, p. 19

Chapter 8

1 Zeeman, 2018
2 History-Biography, 2018
3 Wren, 2005, p. 464
4 Six Sigma Daily, 2018
5 Ahmad, 2018, p. 434
6 Six Sigma Daily, 2018
7 History-Biography, 2018
8 Shingo, 1987, p. 165
9 Cakmakci, 2007, p. 334
10 Liker, 2004, p. 120
11 Cakmakci, 2007, p. 335 Cakmakci, 2007, p. 335
12 Cakmakci, 2007, p. 335
13 Ahmad, 2018, p. 434
14 Cakmakci, 2007, p. 335
15 Luga, 2013, p. 409
16 Ohno, 1988, p. 96
17 Shingo, 1987, p. 165
18 Shingo, 1987, p. xvii
19 Shingo, 1987, p. 166
20 Shingo, 1987, p. 166
21 Shingo, 1987, p. 154
22 Shingo, 1987, p. 154
23 Shingo, 1987, p. 157
24 Womack, 2003, p. 352
25 Shingo, 1987, p. 8
26 Dennis, 2007, p. 96

Chapter 9

1 Crawford-Mason, C., 1980. From an interview with W. Edwards Deming on
   an NBC Special titled "If Japan Can, Why Can't We?" originally aired on
   June 24, 1980.

2 Walton, 1986, p. 15

3 Walton, 1986, p. 3

4 Walton, 1986, p. 6

5 Wren, 2005, p. 462

6 Walton, 1986, p. 7 Walton, 1986, p. 7

7 Walton, 1986, p. 8

8 Deming, 1992 p. 32

9 Wren, 2005, p. 461

10 Walton, 1986, p. 6

11 Wren, 2005, p. 461

12 Wren, 2005, p. 462

13 Walton, 1986, pp. 8-9

14 Walton, 1986, p. 9

15 Wren, 2005, p. 462

16 Walton, 1986 p. 10

17 Walton, 1986 pp. 11-12

18 Wren, 2005, p. 462

19 Wren, 2005, p. 462

20 Walton, 1986, p. 12

21 Walton, 1986, p. 13

22 Walton, 1986, p. 11

23 Walton, 1986, p. 12

24 Liker, 2004, p. 23

25 Walton, 1986 pp. 13-14

26 Walton, 1986, p. 14

27 Walton, 1986, p. 15

28 Walton, 1986, p. 17

29 Walton, 1986, p. 19

30 Wren, 2005, p. 462

31 Wren, 2005, p. 463

32 Wren, 2005, p. 463

33 Walton, 1986, p. 20

34 Meyers, 2002, p. 14

35 Meyers, 2002, p. 14

36 Meyers, 2002, p. 14

37 Keller, 2005, p. 50

38 Gitlow, 2005, p. 693

39 Liker, 2004, p. 264

40 Liker, 2004, p. 264

41 Saier, 2017 p. 147

42 Walton, 1986, pp. 86-87

43 Saier, 2017, p. 147

44 Liker, 2004, p. 246

45 Connelly, 2018, p. 331

46 Connelly, 2018, p. 331

47 Connelly, 2018, p. 332
48 Meyers, 2002, p. 9
49 Meyers, 2002, p. 9
50 Dinero, 2005, p. 51
51 Liker, 1998, p. 58
52 Saier, 2017, p. 144
53 Dinero, 2005, p. 51
54 Womack, 2003, p. 242
55 Liker, 2004, p. 82
56 Wren, 2005, p. 463
57 Wren, 2005, pp. 464-464
58 Dennis, 2007, p. 96
59 Dennis, 2007, p. 96

Chapter 10
1 Ohno, 2013, p. 177
2 Roser, 2015
3 Wren, 2005, p. 464
4 Dinero, 2005, pp. 50-51
5 Liker, 2004, p. 226
6 Liker, 1998, p. xv
7 Liker, 1998, pp. 48-49
8 Liker, 1998, p. 55
9 Graban, 2009, p. 19
10 Womack, 2003, p. 233
11 Womack, 2003, p. 370, note 10
12 Womack, 2003, p. 234
13 Ohno, 1988, pp. 135
14 Liker, 1998, p. 52
15 Ohno, 1988, p. 3
16 Ohno, 1988, p. 20
17 Ohno, 1988, p. 21
18 Ohno, 1988, p. 22
19 Ohno, 1988, p. 9
20 Liker, 1998, p. 48
21 Liker, 2004, pp. 92-93
22 Ohno, 1988, p. 3
23 Ohno, 1988, p. 33
24 Ohno, 1988, p. 25
25 Ohno, 1988, p. 11
26 Womack, 2003, p. 231
27 Womack, 2003, p. 231-232
28 Womack, 2003, p. 232
29 Womack, 2003, p. 232
30 Womack, 19901, p. 56

31 Womack, 1990, p. 53
32 Womack, 2003, p. 233
33 Womack, 2003, p. 234
34 Liker, 2004, p. 21
35 Dennis, 2007, p. 7
36 Liker, 2004, p. 22
37 Liker, 2004, p. 92
38 Dennis, 2007, p. 109
39 Ohno, 1988, p. 11
40 Ohno, 1988, pp. 25-26
41 Liker, 1998, p. 52
42 Liker, 2004, p. 106
43 Ohno, 1988, p. 33
44 Ohno, 1988, p. 111
45 Toyota Motor Corporation, 1998, p. 6
46 Liker, 2004, pp. 106-107
47 Womack, 2003, p. 232
48 Liker, 1988, p. xiv
49 Womack, 2003, p. 242
50 Meyers, 2002, p. 9
51 Ohno, 1988, pp. 34-35
52 Womack, 2003, p. 232
53 Shingo, 1987, pp. 165-166
54 Dennis, 2007, p. 9
55 Womack, 1990, p. 53
56 Womack, 2003, p. 236
57 Dennis, 2007, p. 10
58 Leone, 2014, p. 195
59 Ohno, 1988, p. 33
60 Ohno, 1988, p. 113
61 Ohno, 1988, p. 114
62 Liker, 1998, p. 49
63 Shingo, 1987, p. 173
64 Womack, 2003, p. 243
65 Womack, 2003, p. 221
66 Wren, 2005, p. 465
67 Dennis, 2007, p. 9
68 Jackson, 1996, pp. 4-5
69 Black, 2008, pp. 27 & 30
70 Dennis, 2007, p. 9
71 Womack, 2003, p. 220
72 Liker, 2004, p. 226
73 Liker, 1998, p. 48
74 Womack, 2011, p. 178
75 Womack, 2003, p. 219

76 Liker, 1998, p. 51
77 Ohno, 1988, p. 35
78 Ohno, 1988, p. 69
79 Dennis, 2007, p. 9
80 Dennis, 2007, p. 18
81 Womack, 1990, p. 57
82 Ohno, 1988, p. xi
83 Toyota Motor Corporation, 1998, p. 6
84 Womack, 2003, p. 15
85 Dennis, 2007, pp. 67 & 24
86 Ohno, 1988, p. 41
87 Ohno, 1988, p. 57

Chapter 11

1 Kotter, 1996, p. 5
2 Townsend, 2013, p. 505
3 Dunn, 1987, pp. 238-239
4 Klier, 2009, p. 5
5 Eden, 1996, pp. 513-514
6 Klier, 2009, p. 6
7 Dunn, 1987, p. 235
8 Painter, 2014, p. 190
9 Painter, 2014, p. 193
10 Daito, 2000, p. 139
11 Daito, 2000, pp. 139-140
12 Ohno, 1988, pp. xiii-xiv
13 Liker, 2004, p. 82
14 Catalan Vidal, 2017, p. 5
15 Klier, 2009, p. 7
16 Klier, 2009, pp. 7-8
17 Painter, 2015, p. 198
18 Klier, 2009, p.7
19 Klier, 2009, pp. 8-10
20 Bastos, 2001, p. 431
21 Eden, 1996, p. 518
22 Eden, 1996, p. 519
23 Bastos, 2001, p. 436
24 Bastos, 2001, p. 425
25 Eden, 1996, p. 513
26 Gilewicz, 1976, p. 91
27 Eden, 1996, p. 503
28 Dunn, 1987, p. 243
29 Dunn, 1987, p. 245
30 Eden, 1996, p. 520
31 Dunn, 1987, pp. 245-246

Chapter 12
1 Ford, 1926, p. 244
2 Kiley, 2010
3 Inkpen, 2005, p. 117
4 Kiley, 2010
5 Kiley, 2010
6 Liker, 2004, p. 75
7 Womack, 1990, pp. 82-83
8 Meyers, 2002, p. 9
9 Liker, 2004, pp. 74-75
10 Liker, 2004, p. 75
11 Inkpen, 2005, p. 119
12 Huxley, 2015, p. 135
13 Huxley, 2015, p. 135
14 Inkpen, 2005, p. 118
15 Inkpen, 2005, p. 114
16 Inkpen, 2005, p. 115
17 Liker, 2004, p. 144
18 Kiley, 2010
19 Kiley, 2010

Conclusion
1 De Tocqueville, 2004, p. 15
2 Womack, 1990, p. 13

BIBLIOGRAPHY

Ahmad, Rosmaini, and Mohd Syazwan Faiz Soberi. "Changeover Process Improvement Based on Modified SMED Method and Other Process Improvement Tools Application: An Improvement Project of 5-axis CNC Machine Operation in Advanced Composite Manufacturing Industry." *The International Journal of Advanced Manufacturing Technology* 94, no. 1 (2018): 433–450.

Bastos, Paula. "Inter-firm Collaboration and Learning: The Case of the Japanese Automobile Industry." *Asia Pacific Journal of Management* 18, no. 4 (2001): 423–441.

Black, John R., and David Miller. *The Toyota Way to Healthcare Excellence: Increase Efficiency and Improve Quality with LEAN.* Chicago: Health Administration Press, 2008.

Brucem. "Piggly Wiggly: An Inspiration for the Toyota Way." October 12, 2011. https://xray-delta.com/2011/10/12/piggly-wiggly-an-inspiration-for-the-toyota-way/.

Cakmakci, Mehmet, and Mahmut Kemal Karasu. "Set-up Time Reduction Process and Integrated Predetermined Time System MTM-UAS: A Study of Application in a Large Size Company of Automobile Industry." *The International Journal of Advanced Manufacturing Technology* 33, no. 3–4 (2007): 334–344.

Carden, Art. "Economic Progress and Entrepreneurial Innovation: Case Studies from Memphis." *American Journal of Entrepreneurship* 4, no. 1 (2011): 36–48.

Catalan Vidal, Jordi. "The Stagflation Crisis and the European Automotive Industry, 1973–85." *Business History* 59, no. 1 (2017): 4–34.

"Channing Rice Dooley." Find a Grave Memorial. October 23, 2011. https://www.findagrave.com/memorial/79177533/channing-rice-dooley.

Connelly, Lynne M. "Statistical Process Control." *MedSurg Nursing* 27, no. 5 (2018): 331–333.

Crawford-Mason, Clare, producer *NBC White Paper*, "If Japan Can, Why Can't We?" Aired June 24, 1980 on NBC News. https://www.youtube.com/watch?v=vcG_Pmt_Ny4.

Daito, Eisuke. "Automation and the Organization of Production in the Japanese Automobile Industry: Nissan and Toyota in the 1950s." *Enterprise & Society* 1, no. 1 (2000): 139–178.

Darmody, Peter B.. "Henry L. Gantt and Frederick Tayler: The Pioneers of Scientific Management." *AACE International Transactions* (2007): PS.15.1–PA.15.3.

Defense Health Agency Facilities Division. *DoD space planning criteria: Chapter 450: Sterile processing.* July 1, 2017. https://www.wbdg.org/FFC/DOD/MHSSC/spaceplanning_healthfac_450_2017.pdf.

Deming, W. Edwards. *Out of the Crisis.* Cambridge: MIT press, 2018.

Dennis, Pascal. *Lean Production Simplified: A Plain-Language Guide to the World's Most Powerful Production System* (2nd ed.). New York: Productivity Press, 2007.

De Tocqueville, Alexis. *Democracy in America: A New Translation.* Translated by Arthur Goldhammer. New York: The Library of America, 2004.

Dickens, Milton. "Discussion Method in War Industry." *The Quarterly Journal of Speech* 31, no. 2 (1945): 144–150.

Dinero, Donald. *Training Within Industry: The Foundation of Lean.* Boca Raton, FL: CRC Press, 2005.

Dunn, James A. "Automobiles in International Trade: Regime Change or Persistence?" *International Organization* 41, no. 2 (1987): 225–52.

Eden, Lorraine, and Maureen Appel Molot. "Made in America? The US Auto Industry, 1955–95." *The International Executive* 38, no. 4 (1996): 501–541.

Ford, Henry, and Samuel Crowther. *Today and Tomorrow.* New York: Doubleday, Page & Co, 1926.

Freeman, Mike. *Clarence Saunders & the Founding of Piggly Wiggly: The Rise & Fall of a Memphis Maverick.* Charleston, SC: The History Press, 2013. Kindle.

Gawande, Atul. *The Checklist Manifesto: How to Get Things Right*. New York: Metropolitan Books, 2009.

Gilewicz, P. J., A. C. Gross, and W. W. Ware. "World Motor Vehicle Markets." *Columbia Journal of World Business* 11, no. 1 (1976): 81–93.

Gitlow, Howard S., and David M. Levine. *Six Sigma for Green Belts and Champions: Foundations, DMAIC, Tools, Cases, and Certification*. Upper Saddle River, NJ: Pearson/Prentice Hall, 2005.

Goldratt, E. M., J. Cox, and D. Whitford. *The Goal: A Process of Ongoing Improvement* (3rd Revised). Great Barrington, MA: North River Press, 2004.

Graban, Mark. *Lean Hospitals: Improving Quality, Patient Safety, and Employee Engagement*. New York: Productivity Press, 2009.

Harkins, John E. Review of *Clarence Saunders and the Founding of Piggly Wiggly: The Rise & Fall of a Memphis Maverick* by Mike Freeman. *The West Tennessee Historical Society Papers* 65 (2011):146–149.

History.com Editors. "Interchangeable Parts." Last modified August 21, 2018. https://www.history.com/topics/inventions/interchangeable-parts.

Huxley, Christopher. "Three Decades of Lean Production: Practice, Ideology, and Resistance." *International Journal of Sociology* 45, no. 2 (2015): 133–151.

Inkpen, Andrew C. "Learning through alliances: General Motors and NUMMI." *California Management Review* 47, no. 4 (2005): 114–136.

Iuga, Maria Virginia, and Kifor, Claudiu Vasile. "Lean Manufacturing: The When, the Where, the Who." *Land Forces Academy Review* 18, no. 4 (2013): 404–410.

Jackson, Thomas L., and Karen R. Jones. *Implementing a Lean Management System*. Portland: Productivity Press, 1996.

Jacobs, Ronald L. "Honoring Channing Rice Dooley: Examining the Man and His Contributions." *Human Resource Development International* 5, no. 1 (2002): 131–137.

Keller, Paul (2005). *Six sigma demystified*. New York: McGraw Hill, Inc.

Keys, J. Bernard, Robert A. Wells, and L. Trey Denton. "Japanese Managerial and Organizational Learning." *Thunderbird International Business Review* 40, no. 2 (1998): 119–139.

Kiley, David "Goodbye, NUMMI: How a Plant Changed the
Culture of Car-making." Last modified April 2, 2010, https://www.
popularmechanics.com/cars/a5514/4350856/.

Klier, Thomas. "From Tail Fins to Hybrids: How Detroit Lost Its
Dominance of the US Auto Market." *Economic Perspectives* 33, no.
2 (2009): 2–17.

Kotter, John P. *Leading Change*. Boston: Harvard Business Review
Press, 2012.

Leonardo Group Americas, *"How Piggly Wiggly Revolutionized
Manufacturing: Or, the Quiet Genius of a Milk Rack,"*
Medium (blog), February 4, 2015, https://medium.com/@
LeonardoGroupAmericas/how-piggly-wiggly-revolutionized-
manufacturing-6ae7e76af184.

Leone, Gerard, and Richard D. Rahn. *The Complete Guide to Mixed
Model Line Design: Designing the Perfect Value Stream*. Boulder,
CO: Flow Publishing Inc., 2014.

Liker, Jeffrey K. *Becoming Lean: Inside Stories of US Manufacturers*.
New York: Productivity Press, 1998.

Liker, Jeffrey K. *The Toyota Way: 14 Management Principles from the
World's Greatest Manufacturer*. New York: McGraw-Hill, 2004.

Lopresti, J. (2017, September 14). *Wiggly changed the world of
Lean manufacturing*. https://www.sixsigmadaily.com /kanban-jit-
lean-manufacturing/.

MacArthur, Douglas. *Reminiscences*. Greenwich, CT: Fawcett
Publications, 1964.

Magee, David. *How Toyota Became #1: Leadership Lessons from the
World's Greatest Car Company*. New York: Penguin Group, 2007.

Matsuo, Makoto. "Instructional Skills for On-the-Job Training and
Experiential Learning: An Empirical Study of Japanese Firms."
International Journal of Training and Development 18, no. 4
(2014): 225–240.

Meyers, F. E., & Steward J. R. (2002). *Motion and time study for
Lean manufacturing* (3rd ed.). Columbus, OH: Prentice Hall.

Mirsky, Jeannette, and Allan Nevins. *The World of Eli Whitney*. New
York: Macmillan Company, 1952.

Ohno, Taiichi. *Workplace Management: Special 100th Birthday Edition*.
Translation by Jon Miller. New York: McGraw Hill, 2013.

Ohno, Taiichi. *Toyota Production System: Beyond Large-Scale
Production*. Anonymous Translation. New York: CRC Press, 1988.

Painter, David S. "Oil and Geopolitics: The Oil Crises of the 1970s and the Cold War." *Historical Social Research* 39, no. 4 (2014): 186–208.

Parkes, Aneta. "Lean Management Genesis." *Management* 19, no. 2 (2015): 106–121.

Robinson, Alan G., and Dean M. Schroeder. "Training, Continuous Improvement, and Human Relations: the US TWI Programs and the Japanese Management Style." *California Management Review* 35, no. 2 (1993): 35–57.

Roser, Christoph. "Twenty-Five Years after Ohno—A Look Back." Last modified May 28, 2015. https://www.allaboutlean.com/ohno-25-years/.

Russell, G. (2019). *In search of manufacturing excellence: Toyota & the Piggly Wiggly lesson.* https://www. manexconsulting.com/blog/in-search-of-manufacturing-excellence-toyota-the-piggly-wiggly-lesson/.

Ruttan, Vernon W. *Is War Necessary for Economic Growth?: Military Procurement and Technology Development.* New York: Oxford University Press, 2006.

Saier, Martin Christopher. "Going Back to the Roots of WA Shewhart (and Further) and Introduction of a New CPD Cycle." *International Journal of Managing Projects in Business* 10, no. 1 (2017): 143–166.

Schiffer, Bernd. "[The Pull Principle] People Signing Up for Work." Last modified April 30, 2012. http://agiletrail.com/2012/04/30/people-signing-up-for-work/.

Schur, Joan Brodsky. "Eli Whitney's Patent for the Cotton Gin." Last Modified September 23, 2016. https://www.archives.gov/education/lessons/cotton-gin-patent.

"Shigeo Shingo." History-biography. Accessed June 22, 2018. https://history-biography.com/shigeo-shingo/.

Shingo, Shigeo. *The Sayings of Shigeo Shingo: Key Strategies for Plant Improvement.* Translated by Andrew P. Dillon. New York: Productivity Press, 2018.

Sibaja, Alex. "Jidoka is the Path to Zero Defect." Last modified February 26, 2014. http://alexsibaja.blogspot.com/2014/02/jidoka-is-path-to-zero-defect.html.

Six Sigma Daily Admin. "Who Was Shigeo Shingo and Why Is He Important to Process Improvement?" Last modified November 20, 2018. https://www.sixsigmadaily.com/who-was-shigeo-shingo-and-why-is-he-important-to-process-improvement/.

Soltero, Conrad. "Creating an Adaptable Workforce: Lean Training and Coaching for Improved Environmental Performance." *Environmental Quality Management* 21, no. 1 (2011): 9–22.

Taylor, Frederick Winslow. *The Principles of Scientific Management.* New York: Harper & Brothers Publishers, 1913.

Toma, Sorin-George, and Shinji Naruo. "Total Quality Management and Business Excellence: The Best Practices at Toyota Motor Corporation." *Amfiteatru Economic Journal* 19, no. 45 (2017): 566–580.

Townsend, Susan C. "The 'Miracle'of Car Ownership in Japan's 'Era of High Growth', 1955–73." *Business History* 55, no. 3 (2013): 498–523.

Toyota Industries Corporation. *The story of Sakichi Toyoda.* https://www.toyota-industries.com/company/history/ toyoda_sakichi/.

Toyota Motor Corporation. *The Toyota Production System: Leaner Manufacturing for a Greener Planet.* Tokyo, Japan: Toyota Motor Corporation, 1998.

"Underwriter's Knot: A Two-Strand Crown Knot; Prevents Tension on Terminals." Animated Knots by Grot. Accessed May 29, 2021. https://www.animatedknots.com/underwriters-knot.

Van Vliet, V. (2014). *Sakichi Toyoda.* https://www.toolshero. com/toolsheroes/sakichi-toyoda/.

Wada, Kazuo. "The Fable of the Birth of the Japanese Automobile Industry: A Reconsideration of the Toyoda–Platt Agreement of 1929." *Business History* 48, no. 1 (2006): 90–118.

Walton, Mary. *The Deming Management Method.* New York: Perigree Books, 1986.

Womack, James P. *Gemba Walks.* Cambridge, MA: Lean Enterprise Institute, 2011.

Womack, James P., and Daniel T. Jones. *Lean Thinking: Banish Waste and Create Wealth in Your Corporation*, 2nd ed. New York: Free Press, 2003.

Womack, James P., Daniel T. Jones, and Daniel Roos. *The Machine That Changed the World.* New York: Free Press, 1990.

Woodbury, Robert S. "The Legend of Eli Whitney and Interchangeable parts." *Technology and Culture* 1, no. 3 (1960): 235–253.

Wren, Daniel A. *The History of Management Thought*, 5th ed. Hoboken, NJ: John Wiley & Sons, Inc., 2005.

Wren, Daniel A., and Ronald G. Greenwood. *Management Innovators: The People and Ideas That Have Shaped Modern Business*. New York: Oxford University Press, 1998.

Wukitsch, T. *Rome, Carthage, and the Punic Wars*. http://www.mmdtkw.org/ALRIAncRomUnit4Slides.html.

Zeeman, A. (2018). *Shigeo Shingo*. https://www.toolshero.com / toolsheroes/shigeo-shingo/.

INDEX

Greene, Catherine, 15

Harbour Report, 135
health care, 1, 5, 19, 82, 85, 106, 135
heijunka, 139

Hirano, Hiroyuki, 120
Honda, 130-131

If Japan Can, Why Can't We?, 93, 102
Ikebuchi, Kosuke, 115
Imperial Academy of Inventions, 52
industrial engineering, 27, 81
Industrial Revolution, 11, 13, 15
inventory (waste of), 38-39, 43-45, 53-54,
   75-79, 84, 95, 105, 108-110, 112,
   114-115, 117, 119, 121
inventory, 41, 74, 124, 127

JD Power and Associates, 135
jidoka, 47-49, 51, 52, 80, 88, 106, 108-
   110, 119-120
Job Instruction Training (JIT), 62, 63, 65,
   67, 69-70
Job Methods Training (JMT), 62-64,
   69-71
Job Relations Training (JRT), 62, 64,
   69-70
Jones, Daniel T., 137
Juran, Joseph, 94, 98, 100, 102
Juran Trilogy, 94
just-in-time, 73, 75, 76, 78, 80, 92, 104,
   107, 110-111, 114-115, 117, 119-120,
   127

kaizen, 8, 104
Kane, Mike, 59, 65
kanban(s), 46, 75, 109, 111, 112-115,
   116, 121
Kanzler, Ernest, 117
Keedoozle Corporation, 46
keiretsu, 54, 69
Kotter, John, 125
Korean War, 80, 110, 125
Koromo plant, 108
Koyanagi, Kenichi, 91
Krafcik, John, 137

lean thinking, 139
learn(ing) by doing, 61, 66-67, 73, 80
Lee, Colonel Roswell, 18
Leland, Henry M., 33
lens grinder(s), 59
Liker, Jeffrey, 118, 137
Loom, Type G Automatic, 48-49, 52, 74
MacArthur, Douglas, 54, 69
Machine That Changed the World, The, 137
"Made in Japan," 91
Maruzen, 107
Mellen, Lowell, 69
mistake proofing, 52, 88, 100
motion, 19, 24, 38, 121, 139
motion map, 36
motion study/studies, 27-28, 71, 117
muda, 121
multiplier effect, 61-62, 67, 69-70
mura, 21, 121
muri, 28, 121

Nakajima, Seiichi, 120
Nashua Corporation, 93
NBC, 93, 102
*New Economics for Industry,
   Government, Education, The*, 95
New United Motor Manufacturing, INC
   (NUMMI), 3, 124, 130, 131, 133-137
Nissan, 99, 114, 130-131
nonvalue-added, 11-12, 24, 38-39, 105

Occupational Safety and Health
   Administration (OSHA), 78
Ohno, Taiichi, 11, 40, 45, 50, 70, 83, 98,
   103-111, 114-122, 127, 138, 140
oil crisis/embargo, 53, 115-116, 123-127,
   130
one-piece (flow), 51, 84, 110-111, 119
operating room (OR), 19-20, 37-38, 82,
   85-87, 127-128
Organization of Arab Petroleum Exporting
   Countries (OAPEC), 126-127
overburden(ing), 28, 121
overproduction, 38, 121

perfection, 138-139
persistent/persistence, 3, 6, 9-10, 12, 41,
   46, 117, 140
piece rate, 25, 34

CPSIA information can be obtained
at www.ICGtesting.com
Printed in the USA
BVHW031252280721
613095BV00008B/35/J